D1666258

UB MAGDEBURG
MA9
003 091 546

Elliptic Marching Methods and Domain Decomposition

Patrick J. Roache
Ecodynamics Research Associates, Inc.
Albuquerque, New Mexico

CRC Press
Boca Raton New York London Tokyo

Library of Congress Cataloging-in-Publication Data

Roache, Patrick J.
Elliptic marching methods and domain decomposition / Patrick J. Roache
p. cm -- (Symbolic and numeric computation series)
Includes bibliographical references and index.
ISBN 0-8493-7378-6 (permanent paper)
1. Differential equations, Elliptic--Numerical solutions.
2. Decomposition (Mathematics) I. Title. II. Series.
QA377.R63 1995 95-18775
515'.353--dc20 CIP

International Standard Book Number 0-8493-7378-6
Library of Congress Card Number 95-18775
Printed in the United States of America 1 2 3 4 5 6 7 8 9 0
Printed on acid-free paper

Symbolic and Numeric Computation Series

Edited by Robert Grossman

Published Titles

Computational Mathematics in Engineering and Applied Science, William E. Schiesser
Introduction to Boundary Element Methods, Prem K. Kythe
A Numerical Library in C for Scientists and Engineers, H. T. Lau
Transform Methods for Solving Partial Differential Equations, Dean Duffy

Forthcoming Titles

Numerical Solutions Partial Differential Equations: Problem Solving Using Mathmatica, Victor Grigor'evech Ganzha and Evgenii Valsilevich Vorozhtsov
Symbolic-Numeric Analysis of Dynamic Systems in Engineering, Edwin J. Kreuzer

PREFACE

One of the first things that a student of partial differential equations learns is that one cannot solve elliptic equations by spatial marching. This book is about doing just that.

The spatial marching-out of an elliptic equation from boundary values is unstable (the Hadamard instability) for continuum equations. For discretized equations, which are the principal concerns in modern engineering, science, and applied mathematics, this instability manifests itself as a limitation on the problem size. As we will see herein, there are ways to work around this instability, and the resulting methods have powerful applications to common problems arising in fluid dynamics (external aerodynamics, geophysical flows, flows in porous media, etc.) heat transfer, electrostatics, etc. They are especially well suited for modern Domain Decomposition techniques applied on parallel computer architectures of the MIMD type (multiple-instruction, multiple-data).

Used as direct (noniterative) solvers for *linear* equations, Marching methods are the *only* efficient direct methods capable of handling nonseperable and variable coefficients.

Marching Methods also lead to a class of *iterative* methods for linear problems, some of which converge in two or three applications of the direct algorithm. They also lead naturally to the solution of *nonlinear* steady state problems by *semidirect* methods.

The bulk of this manuscript was completed several years ago, but lay dormant because I came under the impression that marching methods had been largely supplanted by multigrid methods and pre-conditioned conjugate gradient methods. Although marching methods were demonstrably more efficient for particular problems, their quirks and limitations (in regard to problem size, dimensionality, directionality, and robustness) seemed to made them non-competitive for general problems and general user-oriented software.

My opinion has now changed for the better, for two reasons. The first reason is the recent burgeoning of interest in Domain Decomposition techniques, especially as used in the context of MIMD multiprocessor parallel computers. Marching methods are quite naturally suited for this approach. In fact, the stabilizing methods for extending the problem size of marching methods, described in Chapter 3, *are* Domain Decomposition methods that were developed by myself and others before the introduction of the terminology of Domain Decomposition, Schur complement matrix, Schwarz alternating procedure, etc. This book is *not* a "Domain Decomposition" book in the sense that it does not attempt to cover the entire field of Domain Decomposition. However, it *is* a "Domain Decomposition" book in the more limited sense that it covers material that (1) is relevant, sometimes directly and sometimes peripherally, to DD, (2) corrects historical oversights in the mainline DD literature, and (3) exposes readers to some effective alternative DD methods that are still not represented in the mainline DD literature. The fact that the methods are older than the *term* "Domain Decomposition" will not make them any less new to many readers.

The second reason for my improved opinion of marching methods relative to multigrid and PCG methods is my recent close contact with widely acknowledged experts in multigrid and PCG methods who are also working to solve real problems. I am still very impressed with multigrid methods, and I recommend them and indeed *use* them myself for many problems. However, I now realize that multigrid methods have their own quirks and limitations, that the variations necessary for particular problems (contrasted to ideally suited model problems) can adversely affect operation counts and storage penalties (easily by a factor of 2-3 or often more), and that they may require *years* of development to achieve efficiency (or even convergence) for difficult problems (widely varying coefficients, systems of equations, large first derivative terms, Helmholtz terms, etc.). This evaluation applies even more so to PCG methods. As for complexity, a modern algorithm using semicoarsening multigrid (with no restrictions on the numbers of grid points being powers of two) as a pre-conditioner for a conjugate gradient method is many times more complex that any marching method I have ever coded. Also, I now have it on good authority that planar iteration is advisable for some 3D multigrid problems, and that the planar solution algorithm is non-trivial; a marching method is a viable candidate.

The objective herein is efficiency, i.e., speed of solution. Speed is also obtained by increases in computer hardware speed, and some analysts take the position that inefficient methods are all that are required, or at least all that will be required after the next generation of computers debuts. The trouble is that, like "tomorrow", "next" never comes. During my career, some people have been waiting for the "next" generation of computers since 1965. Experience teaches that there will *always* be demands on limited computer resources. Rather than think of the efficiency gain of these algorithms just in terms of computing time, we can think in terms of the radical improvement in problems and reduction in computing power required. These algorithms allow one to progress, for the same use of computing resources, from the coarse grid *qualitative* accuracy to *quantitative* accuracy, from single problem *analysis* to multiple solutions for *design*, from *single* parameter set calculations to *parametric studies*, from *supercomputers* to *microcomputers*, from *batch* runs to *interactive* runs, from *hours* to *minutes*, and sometimes from *2D* to *3D*.

Acknowledgements: It is a pleasure to acknowledge the contributions of professional colleagues and friends who have contributed to my understanding of this and related subjects over many years. These include the following (listed alphabetically): R. E. Banks, R. Bernard, F. G. Blottner, P. Boggs, O. Buneman, R. Coleman, R. T. Davis, F. DeJarnette, D. E. Dietrich, D. Dodson, H. A. Dwyer, E. Gartland, D. Gartling, J. Harris, B. Hulme, M. J. Kascic, D. E. Keyes, P. M. Knupp, T. A. Manteuffel, S. F. McCormick, W. J. Minkowycz, J. D. Morris, C. E. Oliver, T. A. Porsching, S. G. Rubin, J. W. Ruge, A. J. Russo, S. Schaffer, M. Scott, S. Steinberg, C. Temperton, J. P. Thomas. The manuscript word processing was performed by T. Allahdadi, J. Maes and M. Lau. Much of the proofing was performed by J. Maes, S. Brinster, J. D. Morris, and M. Lau, who also converted the figures to computer graphics. I am especially indebted to D. E. Keyes for his scholarly criticism, his sense of style, and his encouragement.

Much of the writing of this manuscript was performed during two short and enjoyable terms as Visiting Professor, one at North Carolina State University, arranged by Prof. F. DeJarnette, and one at University of California at Davis, arranged by good friend Prof. H. A. Dwyer. I have previously summarized much of my own work on marching methods in a series of five articles in the journal *Numerical Heat Transfer*, beginning in Volume 1, Number 1, page 1. I am deeply indebted to Prof. W. J. Minkowycz, the co-founder and editor of NHT, for his encouragement of those articles and this book.

My research in the subject was partially supported by the U.S. Army Research Office, Contract DAAG29-76-C-0018; the Naval Sea Systems Command, General Hydromechanics Research Program Subproject SR 023 01 01 administered by the David W. Taylor Naval Ship Research and Development Center, Contract N00014-76-C-0324; the NASA-Langley Research Center, Contract NAS1-15045, and the U.S. Air Force Office of Scientific Research, Contracts F49620-82-C-0009, F49620-82-C-0064, and F49620-84-C-0079.

TABLE OF CONTENTS

Chapter 3. EXTENDING THE MESH SIZE: DOMAIN DECOMPOSITION

Chapter 4. BANDED APPROXIMATIONS TO INFLUENCE MATRICES

Chapter 5. MARCHING METHODS IN 3D

Chapter 6. PERFORMANCE OF THE 2D GEM CODE

Chapter 7. VECTORIZATION AND PARALLELIZATION

Chapter 8. SEMIDIRECT METHODS FOR NONLINEAR EQUATIONS OF FLUID DYNAMICS

Chapter 9. COMPARISON TO MULTIGRID METHODS

APPENDIX A. MARCHING SCHEMES AND ERROR PROPAGATION FOR VARIOUS DISCRETE LAPLACIANS

APPENDIX B. TRIDIAGONAL ALGORITHM FOR PERIODIC BOUNDARY CONDITIONS

APPENDIX C. GAUSS ELIMINATION AS A DIRECT SOLVER

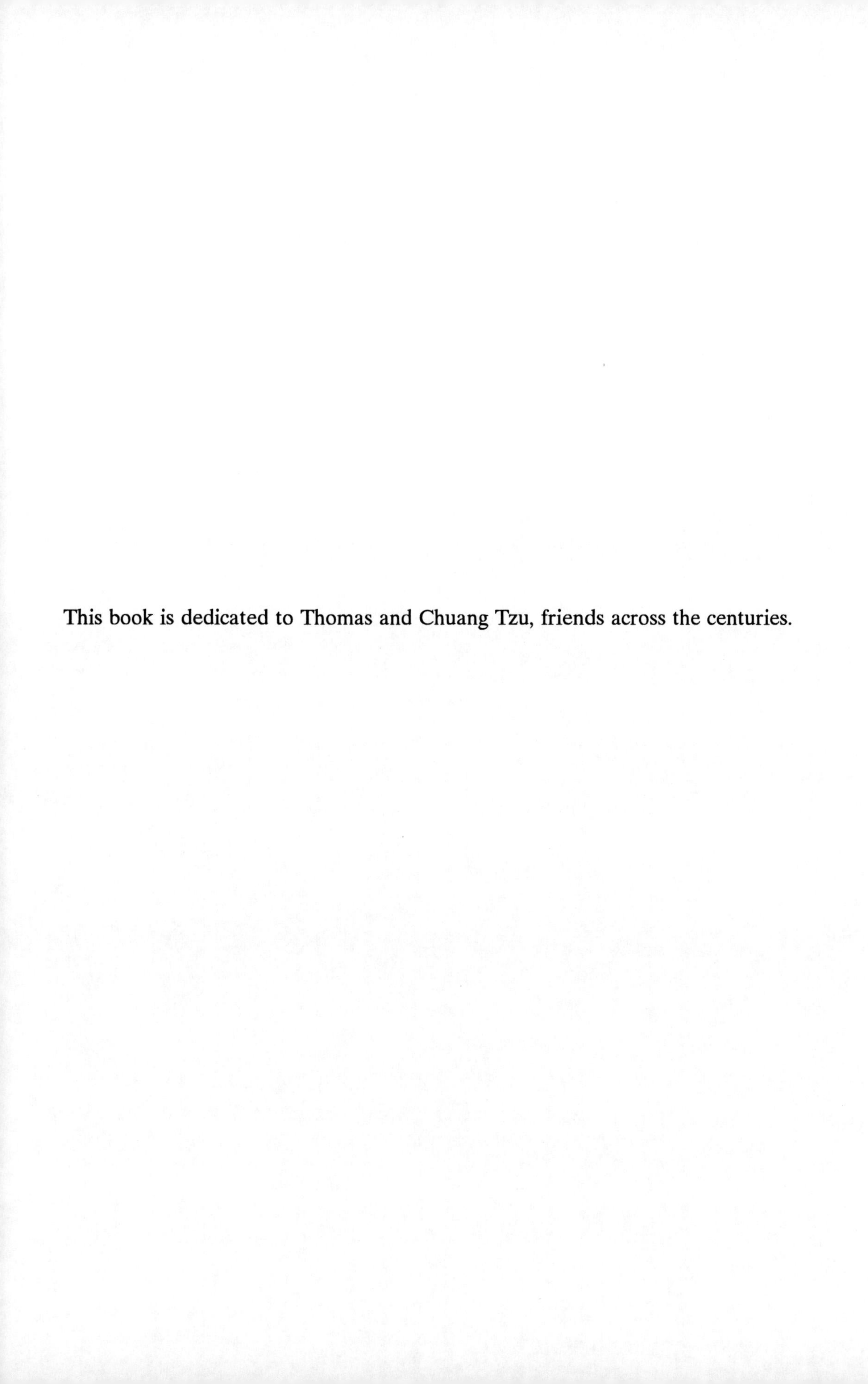

This book is dedicated to Thomas and Chuang Tzu, friends across the centuries.

Chapter 1

BASIC MARCHING METHODS FOR 2D ELLIPTIC PROBLEMS

1.1 Introduction

In this chapter, the history of solving elliptic problems by direct marching methods is reviewed. The basic algorithm is presented, first in 1D (one dimension), then in 2D. Accurate operation counts for initialization and repeat solutions are given and are shown to compare well with other direct methods and with iterative methods. The instability of the marching method is described, and a simple estimate of the resulting mesh size limitation is given. Then the following applications are described: various boundary conditions (Dirichlet, Neumann, Robin, and periodic), irregular meshes, irregular boundaries, interior boundaries, variable coefficient diffusion equations, advection terms (including cell Reynolds number effects and the destabilizing effects of upwind differencing), Helmholtz terms, cross derivatives, turbulence terms, and an expanding grid based on the Fibonacci sequence.

1.1.1 The Impact of Direct Methods

The solution of elliptic partial differential equations by marching methods is an efficient and highly flexible tool for fluid dynamics, heat transfer, electrostatics, and other physical problems that has been under utilized.

Beginning with the very important paper of Hockney [1] in 1965, direct methods for elliptic equations began to supplant the iterative methods such as SOR and ADI. It is difficult to exaggerate the impact of these methods. They are so much more efficient for large problems that problems that had been economically infeasible were now readily calculated. Calculations on 128 x 128 grids are not uncommon, and those on 256 x 256 and larger have been performed. It is still common to introduce research papers with a statement like, "recent advances in computer speed and storage have made possible the large-scale solution of problems in... ." In fact, for large problems, it can be demonstrated that the advances in algorithms like Hockney's method can account for the equivalent advances in speed of one or even two generations of computers.

Hockney's method [1, 2], Buneman's method [3], odd-even reduction [1–5], and others are very powerful but have some limitations. For example, they may be limited to a particular elliptic equation, such as the Poisson equation, or to rectangular boundaries with Dirichlet boundary conditions, or they may depend for their efficiency on the mesh size being a power of 2. They may be extendable, e.g., to nonrectangular boundaries by the *capacitance matrix* technique [2], but at a significant loss of efficiency. Hockney's method is one of the more flexible, as shown by Hockney [2] and LeBail [6], yet neither it nor any of the other fast direct solvers can handle elliptic equations with variable, nonseparable coefficients in each direction or cross derivatives.

Two well-known direct methods for completely general linear algebraic systems are Cramer's rule and Gaussian elimination. Cramer's rule is out of the question even for small systems, since the computing time varies with $n!$ for n equations. For a modest computing mesh of say 20×20 cells, we have $n = 19^2 = 361$, and $n!$ is a literally astronomical number. A more realistic candidate is Gaussian elimination, and this has in fact been used for modest computing meshes. When modern banded Gaussian elimination routines are used, they can even be the method of choice for some 2D problems. Their operation count, storage penalty, etc., are discussed in Appendix C. We note here that they have a large storage penalty compared to the grid for the 2D or 3D problem. The vast difference in the storage requirements arises because the matrix describing the finite-difference equations is very sparse, i.e., most of the elements are zero. The time-like iterative methods do not need to assign storage to these zero elements, but the Gaussian elimination algorithm does since the matrix "fills in" as the elimination proceeds.

1.1.2 Direct Marching Methods

The direct marching methods have none of these shortcomings but have another one: they are unstable. This has been widely known, and the methods have been widely dismissed as useless. What is not so widely known is that they are eminently usable for an important class of problems, those with a cell aspect ratio $\Delta x/\Delta y > 1$. This condition is very common in practical fluid dynamics and heat transfer problems, especially in aerodynamics, and the marching methods can find excellent applications there. Furthermore, they can be stabilized even for $\Delta x \leq \Delta y$, although at a significant loss of efficiency. They are naturally adapted to Domain Decomposition techniques, and in fact, several of the stabilizing methods *are* Domain Decomposition methods. They also have a property that is of considerable mathematical interest to the area of *computational complexity*: after some preprocessing, they have an operation count that is "optimal", i.e., merely proportional to the number of unknowns.

1.1.3 History of Marching Methods

The history of marching methods for elliptic partial differential equations is difficult to trace for two reasons. First, the basic idea is very old. Second, when linear equations are being considered, it is difficult to distinguish methods from algorithms.

The basic idea is to solve a boundary value problem, with split boundary conditions, by guessing the missing conditions at one boundary and "marching" (or "shooting") the solution, as an initial-value problem, to the second boundary. The resulting final values at the end of the march are compared with the desired boundary values, and on that basis the guessed conditions are corrected and the march is repeated for the final correct solution. (Alternately, different marched solutions may be stored, and final correct solutions can be constructed by a linear combination of these in the *superposition* methods.) For nonlinear equations, the corrections are approximate and the shooting methods are iterative. For linear equations, the correction can be exact and only two marches are required to obtain the solution. These methods are commonly used for ordinary differential equations. For elliptic partial differential equations, they are unstable. The *continuum* equations are unstable, as shown by Hadamard [7]; the solution away from the boundary becomes sensitive to infinitesimal perturbations in initial data, and the problem is said to be "ill-posed in the sense of Hadamard" [7]. This is reflected in the finite-difference equations, which fail as $\Delta x \to 0$. However, for finite Δx, solutions may be obtained, as we shall see.

All marching methods share these features, so one may ask whether they can fairly be called distinct *methods* at all. This is not an easy question to dismiss. Since we are dealing with linear equations, and since all the methods are "direct," i.e., they produce the algebraically exact answer in a finite number of steps, it is assured that all the methods will be related to one another via linear algebra. Yet they may use different procedures and formulas, have different operation counts, and have different susceptibilities to computer round-off error. In these instances, we can say at least that the *algorithms* are distinct.

The particular marching algorithm and notation used in this work is one that I originally called the EVP method [8–10]. The earliest journal reference to a marching method for discretized linear partial differential equations that I have found was presented as a hand calculation, obviously suitable only for coarse grid solutions, in the 1957 paper by Ishizaki [11] in the bulletin of the Disaster Prevention Research Institute at Kyoto University. In a more accessible reference, Bickley and McNamee [12] discussed in a general way the instability of marching elliptic equations. The earliest publication of essentially what I later called the EVP method was published by Booy [13] in 1966. (I am indebted to Dr. R. Coleman for providing me with this reference as well as [18].) The GSM method of Hirota et al. [14], published months before [8], is almost, but not quite, identical to the EVP algorithm (see below). The major unique contribution of [8, 9] appears to be the explanation of the dominant importance of the cell aspect ratio on the practicality of the method.

Other papers and their methods deserve comment. The method of Schechter [15, 5] is a marching method but has a significantly different matrix description and operation count than does the EVP algorithm. The method of Lucey and Hansen [16] was apparently one of the first to include a stabilizing scheme, but is not applicable to irregular regions. Coleman's method [17] is a stabilized version of Booy's method [13], yet is obviously distinct algorithmically, since it cannot treat first derivatives in the march direction while other algorithms can. (With a modification, it can treat constant-coefficient first derivatives in the march direction.) Another stable method is the *dynamic programming* of Angel and Bellman [18], which converts the elliptic problem to two initial-value problems with a variational context. However, dynamic programming is much less efficient than the presently considered marching method in both computer time and storage requirements. References to other early marching methods can be found in the work of Dorr [5, 19] and Bank and Rose [20], to which we add the following. Dietrich et al. [21] used the GSM method on partitioned blocks to solve subproblems directly and then iteratively solved for the boundary values of the subproblems; we will sketch their BIR method in Chapter 3. Yee [22] solved the Poisson equation on the surface of a sphere using a marching method but with a distinctive algorithm utilizing a Fourier transform to solve for the missing initial condition.

The work of Bank and Rose [20, 23–25] brought the marching method to a new level of sophistication in the matrix description, in stabilization, and in the analytic generation of the influence coefficient matrix (see the section on operation count below) by way of Chebychev polynomials for constant coefficient equations (see also Dorr [19]). The present chapter discusses applications and extensions of what Bank and Rose call "simple marching" and is a separate line of development from theirs. The mathematical sophistication of the Bank and Rose papers is high in comparison to the present work, which utilizes only the most basic concepts of linear algebra. It is hoped that the present work will complement their work and will serve to extend the application of these powerful marching methods for elliptic equations.

1.2.1 The Marching Method in 1D

We are concerned with obtaining numerical solutions of discretized linear elliptic equations, such as the Poisson equation

$$\nabla^2 \Psi = \zeta \tag{1.2.1}$$

in two or three dimensions (which we abbreviate as 2D and 3D). However, the marching method is most easily introduced by way of the 1D problem (which can be solved by the more conventional Thomas tridiagonal algorithm, e.g., as in Appendix A of Ref. 10).

As a reference elliptic problem in 1D, we consider the one-dimensional Cartesian form of Eq. (1.2.1), and consider the problem with Dirichlet two-point boundary conditions.

$$d^2\Psi/dy^2 = \zeta \tag{1.2.2}$$

$$\Psi(0) = a, \qquad \Psi(1) = b \tag{1.2.3}$$

Using the second-order accurate centered difference approximation gives

$$\Psi_{j+1} - 2\Psi_j + \Psi_{j-1} = \Delta y^2\, \zeta_j \tag{1.2.4}$$

where j runs from 1 to J, and $\Delta y = 1/(J-1)$. The boundary conditions are

$$\Psi_1 = a \quad , \quad \Psi_J = b \tag{1.2.5}$$

We now pick an arbitrary value of Ψ'_2, where the prime denotes a provisional value, say $\Psi'_2 = \Psi_1 = a$. This Ψ'_2 is in error from the true value Ψ_2 by the error $\mathbf{e}$, that is,

$$\Psi_2 = \Psi'_2 + \mathbf{e} \tag{1.2.6}$$

The remaining provisional values up through J are now marched out in one sweep, starting at $j = 3$, by rearrangement of Eq. (1.2.4).

$$\Psi'_{j+1} = \Delta y^2 \zeta_j + 2\Psi'_j - \Psi'_{j-1} \tag{1.2.7}$$

These provisional values Ψ'_j are in error by $\mathbf{e}_j$, that is,

$$\Psi_j = \Psi'_j + \mathbf{e}_j \tag{1.2.8}$$

Substituting Eq. (1.2.8) into Eq. (1.2.4) and using Eq. (1.2.7), we obtain the recursion relation for the error propagation as

$$\mathbf{e}_{j+1} = 2\mathbf{e}_j - \mathbf{e}_{j-1} \tag{1.2.9}$$

which is seen to be independent of the non-homogeneous term ζ_j. For the presently considered boundary conditions, we have $\mathbf{e}_1 = 0$ and $\mathbf{e}_2 = \mathbf{e}$. Applying Eq. (1.2.9) recursively gives

$$\begin{aligned}
\mathbf{e}_1 &= 0 \\
\mathbf{e}_2 &= \mathbf{e} \\
\mathbf{e}_3 &= 2\mathbf{e}_2 - \mathbf{e}_1 = 2\mathbf{e} \\
\mathbf{e}_4 &= 2\mathbf{e}_3 - \mathbf{e}_2 = 2(2\mathbf{e}) - \mathbf{e} = 3\mathbf{e} \\
\mathbf{e}_5 &= 2\mathbf{e}_4 - \mathbf{e}_3 = 2(3\mathbf{e}) - 2\mathbf{e} = 4\mathbf{e} \\
&\vdots \\
\mathbf{e}_j &= (j-1)\cdot\mathbf{e} \quad , \quad \mathbf{e}_J = (J-1)\cdot\mathbf{e}
\end{aligned} \tag{1.2.10}$$

In this simple system, the term (J-1) is the value of the "influence coefficient", giving the influence of a unit perturbation (or error) at j = 2 on the value at j = J.

At the end of the first sweep, the unit error **e** is calculated from the known boundary value b and Eqs. (1.2.8) and (1.2.10) as

$$\mathbf{e} = \frac{b - \Psi'_J}{J - 1} \tag{1.2.11}$$

With **e** so determined, the provisional values are now corrected to the final values in a second march, using Eqs. (1.2.10) and (1.2.8)

$$\Psi_j = \Psi'_j + (j-1)\cdot\mathbf{e} \tag{1.2.12}$$

Thus, the final answer is obtained directly in two successive (forward) marches of the elliptic equation.

There is superficial resemblance between this 1D marching algorithm and the usual (Thomas) tridiagonal algorithm, but in fact the algorithms are quite distinct. In the usual tridiagonal algorithm, two "sweeps" are also used, but with one in each direction, and the algebra is different. The distinction manifests itself strongly when one considers the general coefficient problem (e.g., see Appendix B of [10]) and allows for interior boundary values, i.e. the degenerate or "dead cell" problem for which a fixed value ψ = c occurs at some interior value j = j_c. The resulting matrix could be partitioned into two separate

Dirichlet problems and solved either by marching or the tridiagonal algorithms. However, the tridiagonal algorithm can also solve the original (non-partitioned) matrix problem, whereas the marching algorithm will fail with a divide by zero. (The error e_j does not propagate through the interior boundary $j = j_c$. See also Section 1.3.8.) Also, the tridiagonal algorithm requires one or two auxiliary storage vectors, whereas the marching method does not.

There also is an obvious connection between this method and the method of superposition, in which any solution is obtained as a linear combination of a particular solution to the non-homogeneous equation with non-homogeneous boundary conditions (the Ψ' solution) and a homogeneous solution (the **e** solution). In fact, some users of marching methods, e.g., Dietrich [21], actually use superposition. For complex equations in 1D it involves less algebraic operations, but requires more storage.

Since the recursion relation [Eq. (1.2.9)] for the error propagation is linear in j (for the Poisson equation, at least), there is no practical danger of generating excessively large $\Psi'(j)$ and thereby destroying accuracy due to machine round-off error. Unfortunately, the multidimensional version of this method is not free from this shortcoming.

1.2.2 The Reference 2D Problem

As a reference elliptic problem in 2D, we consider the Poisson equation in Cartesian coordinates.

$$\nabla^2\Psi \equiv \frac{\partial^2\Psi}{\partial x^2} + \frac{\partial^2\Psi}{\partial y^2} = \zeta(x,y) \tag{1.2.13}$$

We consider a rectangular domain (Fig. 1.2.1) of dimensions X and Y, with constant Δx and Δy, using the usual second-order accurate, 5-point difference analog of the Laplacian operator. With

$$\Delta x = \frac{X}{I-1} \qquad \Delta y = \frac{Y}{J-1} \tag{1.2.14}$$

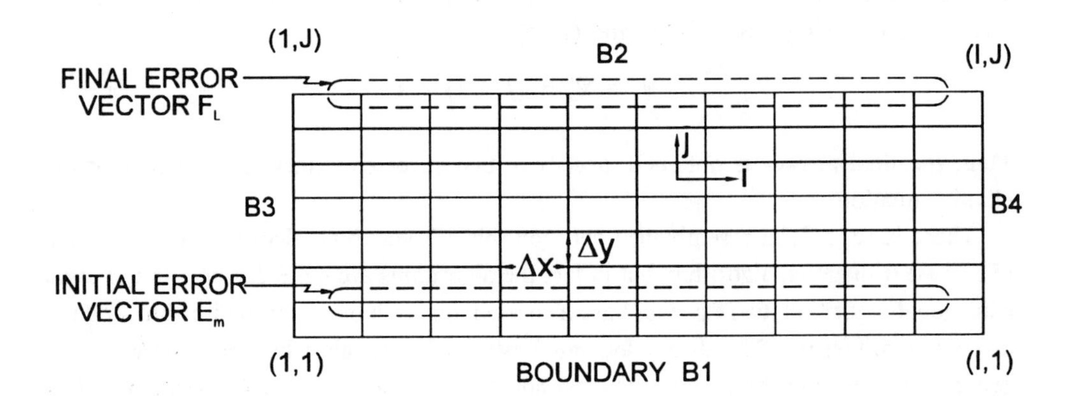

Figure 1.2.1. Geometry of the Reference Problem for the Marching Method.

we have

$$\frac{\Psi_{i+1,j} - 2\Psi_{i,j} + \Psi_{i-1,j}}{\Delta x^2} + \frac{\Psi_{i,j+1} - 2\Psi_{i,j} + \Psi_{i,j-1}}{\Delta y^2} = \zeta_{ij} \tag{1.2.15}$$

For concreteness, we first consider Dirichlet boundary conditions at all boundaries, with $\Psi_{i,1}$, $\Psi_{i,J}$, $\Psi_{1,j}$, and $\Psi_{I,j}$ specified along boundaries B1, B2, B3, and B4, respectively.

We now pick an arbitrary vector of provisional values $\Psi'_{i,2}$ just inside boundary B1, say $\Psi'_{i,2} = \Psi_{i,1}$. This $\Psi'_{i,2}$ is in error by the *error* vector $\mathbf{e}_{i,2}$:

$$\Psi_{i,2} = \Psi'_{i,2} + \mathbf{e}_{i,2} \tag{1.2.16}$$

With $\Psi'_{i,2}$ so chosen, the remaining provisional values for $2 \leq i \leq (I - 1)$ and j up to J (boundary B2) are calculated in one march, starting at $(i, 3)$, by rearrangement of Eq. (12.15),

$$\Psi'_{i,j+1} = \Delta y^2 \zeta_{i,j} + (2 + 2\alpha)\Psi'_{i,j} - \alpha\left(\Psi'_{i+1,j} + \Psi'_{i-1,j}\right) - \Psi'_{i,j-1} \tag{1.2.17}$$

where $\alpha = (\Delta y/\Delta x)^2$. The correct boundary values $\Psi_{1,j}$ at B3 and $\Psi_{I,j}$ at B4 are used in Eq. (1.2.17) when needed. The error propagation equation is then

$$\mathbf{e}_{i,j+1} = (2 + 2\alpha)\mathbf{e}_{i,j} - \alpha\left(\mathbf{e}_{i+1,j} + \mathbf{e}_{i-1,j}\right) - \mathbf{e}_{i,j-1} \tag{1.2.18}$$

with boundary values along B1, B3, and B4 of

$$\mathbf{e}_{i,1} = \mathbf{e}_{1,j} = \mathbf{e}_{I,j} = 0 \tag{1.2.19}$$

After the first march of $\Psi'_{i,j}$, the values of the final error vector $\mathbf{e}_{i,J}$ are calculated from

$$\mathbf{e}_{i,J} = \Psi_{i,J} - \Psi'_{i,J} \tag{1.2.20}$$

where $\Psi_{i,J}$ is the known boundary value vector along B2. From Eq. (1.2.18) a linear relation may be established, allowing the solution for $\mathbf{e}_{i,2}$ from $\mathbf{e}_{i,J}$. The correct values of $\Psi_{i,2}$ are taken solved from Eq. (1.2.16), and a second march using the recursion relation, Eq. (1.2.17) (with Ψ replacing Ψ'), establishes the final solution.

To establish the linear relation allowing the solution for $\mathbf{e}_{i,2}$ in terms of $\mathbf{e}_{i,J}$, it is convenient to introduce two vectors, shown in Fig. 1.2.1. The final error vector is defined as $\mathbf{F}_l = \mathbf{e}_{i,J}$, where l $(= i - 1)$ runs from 1 to $(I - 2)$. The initial error vector is defined as $\mathbf{E}_m = \mathbf{e}_{i,2}$, where m $(= i - 1)$ also runs from 1 to $(I - 2)$. The influence coefficient matrix $C = [C_{lm}]$ is defined as

$$\mathbf{F}_l = C_{lm}\mathbf{E}_m \tag{1.2.21}$$

In the one-dimensional case, the error propagation is linear in j, and C degenerates to the scalar $(J - 1)$ in Eq. (1.2.10). In two dimensions, no really simple equation exists for

C_{lm}. (Bank and Rose [20, 23–25] have developed such relations for this reference problem in terms of Chebychev polynomials, but these are not applicable to the more general problems we will be discussing.) The matrix C is established prior to the solution of a particular problem by the following process. Taking a particular value m_1 of m, set $\mathbf{E}_{m_1} = \mathbf{e}_{m_1+1,2} = 1$, with all other $\mathbf{E}_m = 0$. Then the propagation of the error vector $\mathbf{E}$ into $\mathbf{e}_{i,j}$ is calculated by application of Eqs. (1.2.18, 1.2.19), and the resulting final error vector is $\mathbf{F}_l = \mathbf{e}_{i,J}$, where $l = (i - 1)$ runs from 1 to $(I - 2)$. This determines the m_1-th column of C as

$$C_{l,m_1} = \mathbf{F}_l \tag{1.2.22}$$

Repeating the generation of $E_{m_1} = 1$, with all other $\mathbf{E}_m = 0$ and for m_1 ranging from 1 to $(I - 2)$, fills in the influence coefficient matrix C. Finally, to solve for $\mathbf{e}_{i,2}$ we solve Eq. (1.2.21) using direct Gaussian elimination and formally obtain

$$\mathbf{E}_m = C_{ml}^{-1} \mathbf{F}_l \tag{1.2.23}$$

$$\mathbf{e}_{m+1,2} = C_{ml}^{-1} \mathbf{e}_{l+1,J} \tag{1.2.24}$$

In practice, we do not actually obtain C^{-1}, but just solve Eq. (1.2.21) for $\mathbf{E}_m$ by the Gaussian elimination on the first call, and retain the LU triangular decomposition coefficients for later calls. (For a clear exposition of the LU decomposition and other practical aspects of linear algebra, see Blum [26], Dahlquist and Björck [27], Forsythe, Malcolm and Moler [28], or Golub and Van Loan [50].)

In the next three sections, the performance of the marching methods is described both in regard to "operation counts", i.e., computer time requirements, and in regard to stability, i.e., problem size limitations. The reader interested only in applications can simply scan these sections to get a feel for the results. The only result that is critical to an application is equation (1.2.38) which gives the problem size limitation. (This will be further extended by techniques to be described in Chapter 3.)

1.2.3 Operation Counts As An Index of Merit

The efficiency of the marching method can be compared to other methods on the basis of "operation counts", denoted by θ. The operation count θ provides a single comparative index for different methods, and is very useful in that regard. However, it is important to realize at the outset that this procedure has some important drawbacks.

1. There is little uniformity in counting operations, with some authors counting only multiplications and division, others including additions and subtractions, and sometimes even replacements; the result is that the coefficient of the operation count is ambiguous.
2. We do not count the computer time required to do subscript indexing in loops (e.g., Fortran DO loops). This can be significant, especially in 3D problems, but

cannot practically be accounted for here because it is highly dependent on computer hardware, software, and programming techniques.

3. The true relative advantage of methods will be computer dependent even on "conventional" scalar computers, since the relative speeds of the arithmetic and indexing operations are computer (and software) dependent.
4. On advanced computers that use more sophisticated architecture (such as parallel processing, pipeline processing, vector processing, and even stack-loop operations) other speed considerations may well outweigh the simple operation count. These other considerations include the "paging" requirements from slow memory storage, "vectorizability" of the algorithm, and even the occurence of "IF" statements (which close down a pipeline stream).
5. The counts are often only made for $\Delta x = \Delta y$, which can introduce a slight ambiguity, and for equal number of mesh points in each direction, which can introduce a serious ambiguity in methods that have a strong directional quality. This is especially the case for marching methods, and to a lesser extent for Hockney's method.
6. Most authors keep only the highest ordered terms, so that the index θ is only meaningful for asymptotically large problems.
7. Comparisons with iterative methods (such as SOR, ADI, PCG and multigrid) are necessarily based on an arbitrarily specified iteration convergence criterion, which can substantially affect comparisons.
8. The operation count θ is often taken as the sole index of merit of a method; in fact, flexibility, ease of programming and storage requirements may be equally important considerations.

1.2.4 Operation Counts for the Reference 2D Problem

The comparison is based on solving the Poisson equation in a rectangular domain; however, we again emphasize that the method is applicable to more general problems. (The values for θ given here are more accurate than those in the original study [9], especially in regard to the work for Gaussian elimination.)

Consider first the operation count for the initialization. Each application of error propagation equation (1.2.18), which is homogeneous, requires two multiplications and three additions or subtractions; we write $2*, 3\pm$. [The coefficient $(2 + 2\alpha)$ is prestored, i.e., calculated just once, outside the DO loop of the algorithm.] Applied to each of $L \times M$ points [where $L = (I - 2)$ and $M = (J - 2)$], this gives an operation count of

$$\theta_1 = (2LM)*, \ (3LM)\pm \tag{1.2.25}$$

(For $\Delta x = \Delta y$ we have $\alpha = 1$, which gives one less multiplication in Eq. (1.2.18), but $\Delta x = \Delta y$ is not the best situation in which to use the method; also, we want more generality.) To establish the influence coefficient matrix C, we need L of these marches, which would give $\theta_2 = (2L^2M)*$ and $(3L^2M)\pm$. However, some of these applications of Eq. (1.2.18) to $L \times M$ points are null calculations; after we set $\mathbf{E}_m \equiv \mathbf{e}_{m_1+1,2} = 1$ and march, we find $\mathbf{e}_{i,j} \neq 0$ only for $(m_1 + 1 - j) < i < (m_1 + 1 + j)$. The other nodes in the "zone of silence" of Fig. 1.2.2 give null calculations. For $L \leq M$, there are $L^2(L-1)/3 \cong L^3 3$ of these null calculations. If these are avoided, we obtain reduction by a factor λ, giving

$$\theta_2 = (2\lambda L^2 M)*,\ (3\lambda L^2 M)\pm \tag{1.2.26a}$$

$$\lambda = 1 - \frac{1}{3}\frac{L}{M}, \qquad L \leq M \tag{1.2.26b}$$

(This reduction factor for null calculations may be verified by the reader following Fig. 1.2.2 for all m_1, in a somewhat clumsy accounting.) For the case of $L \gg M$, we have $\lambda \rightarrow (M+2)/L$. The solution of the $L \times L$ influence coefficient matrix C is accomplished by LU decomposition, which is retained for later repeat solutions. According to Blum [26], dense LU decomposition for n equations requires

$$\theta_{LU} = \left[\frac{1}{3}n^3 + \frac{1}{2}n^2\right]*,\ \left[\frac{1}{3}n^3 + \frac{1}{2}n^2\right]\pm,\ \left[\frac{1}{2}n^2\right]\div \tag{1.2.27}$$

In the present case, $n = L$, giving

$$\theta_3 = \left[\frac{1}{3}L^3 + \frac{1}{2}L^2\right] * \text{ and } \pm,\ \left[\frac{1}{2}L^2\right]\div \tag{1.2.28}$$

plus some trms of $O(L)$. Thus, the initialization procedure requires $\theta_2 + \theta_3$ operations, or

$$\theta_{init} = \left[2\lambda L^2 M + \frac{1}{3}L^3 + \frac{1}{2}L^2\right]*, \\ \left[3\lambda L^2 M + \frac{1}{3}L^3 + \frac{1}{2}L^2\right]\pm,\ \left[\frac{1}{2}L^2\right]\div \tag{1.2.29}$$

The operation count for repeat solutions involves two marches of Eq. (1.2.17) and the solution for $\mathbf{e}_{i,2}$. For constant Δx and Δy, careful programming of the physical problem can usually eliminate the multiplication of Δy^2 and $\zeta_{i,j}$ in Eq. (1.2.17), i.e., the FORTRAN variable represents the product. Thus, each application of Eq. (1.2.17) requires $2*$ and $4\pm$ at each of $L \times M$ points (or one less $*$ for $\Delta x = \Delta y$). For two marches, this gives

$$\theta_4 = (4LM)*,\ (8LM)\pm \tag{1.2.30}$$

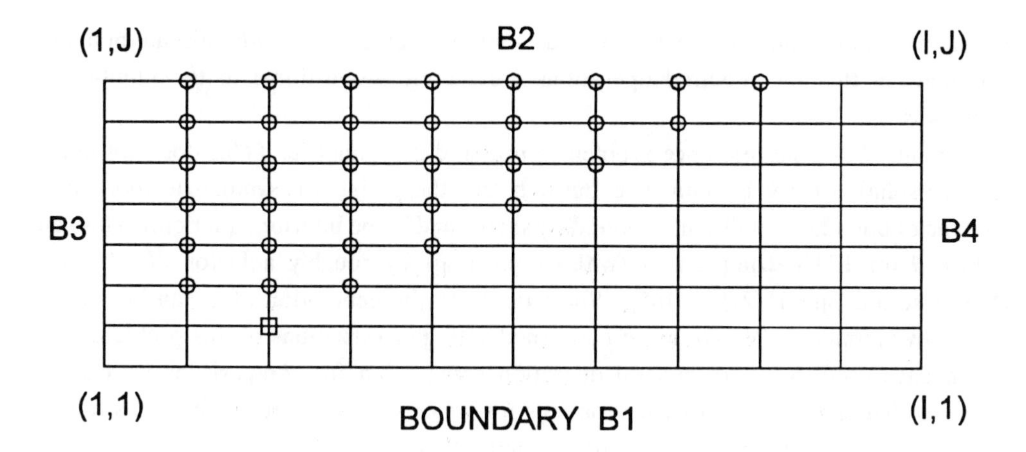

Figure 1.2.2. Null calculations for the march of a 2D 5-point stencil with Dirichlet boundary conditions for the m1 column of Cℓm. ◻ denotes the unit error vector, and O denotes resulting non-zero error values. Unmarked nodes (+) are in the "zone of silence" of the march, requiring no arithmetic operations during the initiation stage of the marching method.

With the LU decomposition previously initialized, the "backsolutions" for $\mathbf{e}_{i,2}$ require (see Dahlquist and Björck [27], p. 154)

$$\theta_{\text{BS}} = (n^2)*,\ (n^2)\pm \tag{1.2.31}$$

In the present case, $n = L$, giving

$$\theta_5 = (L^2)*,\ (L^2)\pm \tag{1.2.32}$$

plus terms of $O(L)$. The solution for $\Psi_{i,2}$ from $\Psi'_{i,2}$ and $\mathbf{e}_{i,2}$ requires $(L)\pm$, which we likewise neglect. Thus, the repeat solutions require $(\theta_4 + \theta_5)$ operations, or

$$\theta_{\text{rep}} = (4LM + L^2)*,(8LM + L^2)\pm \tag{1.2.33}$$

These operation counts in Eq. (1.2.29) and (1.2.33) are accurate but difficult to use as a basis of comparison with other methods. Following Dorr [5] and others, we assume $L = M$ (but not $\Delta x = \Delta y$) and count each $*$, $\div$, or $\pm$ as an operation, neglecting the differences in computer execution times for these different arithmetic operations. For $L = M$, Eq. (1.2.26b) gives $\lambda = 2/3$, and we obtain

$$\theta_{\text{init}} = 4M^3 + \frac{3}{2}M^2 \tag{1.2.34a}$$

$$\theta_{\text{rep}} = 14M^2 \tag{1.2.34b}$$

From these equations we see that repeat solutions for $L = M$ with this method are "optimal" in that the required operations are merely proportional to the number of unknowns.

Actually, C for the reference problem considered is symmetric. (This does not imply that the solution for Ψ is symmetric, but only that the problem is symmetric about *ic* in the type of boundary condition, in boundary shape, and in the interior equation.) This can reduce θ for LU decomposition (without pivoting) by roughly a factor of 1/2 (see Dahlquist and Björck [27], p. 162). Similarly, θ for the generation of C can be almost halved by symmetry. We are more concerned with generalizations of this problem, and so we have used the more general operation count. But for comparisons with other methods that cannot readily treat problems for which C is asymmetric, it seems fair to incorporate symmetry into our operation count. This gives roughly

$$\theta_{init,sym} \cong 2M^3 \tag{1.2.35}$$

The best direct methods given by Dorr [5] have an operation count of the form

$$\theta_{tot} = aM^2 \ln M \tag{1.2.36}$$

where $a = 9/2$ for the odd-even reduction method and the nearly equal value of $a = 5$ for the Fourier series method. In [19], the form of Eq. (1.2.34) giving $\theta = O(M^2 \ln M)$ depends on $(M + 1)$ being a power of 2, which is irrelevant to the marching algorithm; for general M, these methods give $\theta = O(M^3)$. However, Sweet [29] extended the Buneman odd-even reduction algorithm to arbitrary M and still attained $\theta = O(M^2 \ln M)$.

Neither of these direct methods exhibits a significant penalty for initialization. Compared with the Fourier series method, the present algorithm has a larger operation count for the initial solution by a factor of about $(2/5\ M)/\ln M > 1$ but a smaller operation count for repeat solutions by a factor of about $14/(5 \ln M)$ (which is <1 for $M \geq 17$). For practical mesh sizes ($M = 32 - 128$) this means that 20–25 repeat solutions are required before the marching method will show greater speed. However, further advantage accrues if an irregular boundary is considered. For example, if the upper boundary is not along constant j, the application of the Fourier series methods (by the point charge or capacity matrix technique applied in the most straightforward way [2]) requires M additional solutions of a Poisson equation for the initial solution plus a matrix solution similar to the marching method itself. If only the M Poisson solutions are considered, this gives an operation count for the initial solution by a Fourier series method of $\theta = 5M^3 \ln M$. The operation count for the marching method does not change, as we will see in the next section, making it better than the Fourier series methods by a factor of about $2/(5 \ln M)$ even for the initial solution. If $(M + 1)$ is not restricted to powers of 2, the comparison is even more favorable to the marching method. Note, however, that Proskurowski and Widlund [30] have shown that the irregular region may be imbedded in a larger problem with periodic boundary conditions in one direction; the periodicity makes the capacity matrix circulant for the constant coefficient Poisson equation so that it can be established by only one solution of the Poisson equation plus shifting operations.

Comparisons of any of these direct methods with iterative methods such as SOR are overwhelming for repeat calculations. The SOR method (e.g., see Roache [10], p. 118) requires 3 * and 6 $\pm$ per mesh point, or

$$\theta_{SOR} = (3LM)*, (6LM)\pm = 9LM \tag{1.2.37}$$

operations per sweep, which is 50% more than a single march from Eq. (1.2.30). Comparing this method for $L = M$ with Eq. (1.2.34a) we see that marching *initialization* requires $4M/9$ times the operations for a single SOR sweep. Since it requires more than M iterations of SOR to obtain a dependable solution (e.g., see Hockney [2]), marching initializes in less operations than one SOR solution. (This is without even considering the difficult problem of obtaining near-optimal under-relaxation coefficients for SOR in complex geometries with general combinations of boundary conditions.) Comparing Eq. (1.2.37) with Eq. (1.2.34b) we see that *repeat* solutions for marching require less operations than *two* SOR iterations. For large problems, this can easily mean 2 orders of magnitude savings over SOR. (Dorr [5] estimated the operation count required to obtain solutions by the iterative methods to be $O(M^3 \ln M)$ for SOR and $O(M^2 \ln M)$ for ADI. See also [51].) Comparisons with multigrid methods are the subject of Chapter 9.

There is a storage penalty involved for the marching method, since C is an additional $L \times L$ matrix, roughly equivalent to the storage for Ψ itself. If ζ is also stored, as in typical codes, this means a 50% storage penalty for the marching method. (If C is symmetric, the storage penalty can be reduced by 1/2.) In some linear algebra packages, the Gaussian elimination routine [31] requires an additional work storage area of approximately the same size as C. This work array is required if the solver reduces the round-off effects by a "cleanup" iteration, which may or may not be necessary.

This storage penalty is much smaller than the storage penalty for the dynamic programming method [18], and the operation count is also much better. Dynamic programming requires $O(M^4)$ operations for initialization and $O(M^3)$ for repeat solutions, compared with $O(M^3)$ and $O(M^2)$, respectively, for the marching method.

We emphasize that although the operation count comparisons with other direct methods for the model problem are favorable to the marching method, equally important advantages are its flexibility and simplicity. These are to be contrasted with its field size limitations because of the error propagation characteristics, which we now consider.

1.2.5 Error Propagation Characteristics for the Reference 2D Problem

There is a tendency for C to become ill-conditioned as I becomes large. In our experience on long word-length computers such as a Cray, this becomes a problem for $I > 100$. It can be improved by including double-precision steps in the Gaussian elimination routine [31], which steps are costly on a Cray but cheap on a workstation. Usually, the more stringent limitation results from the error propagation as J gets large.

Unlike the recursion relation Eq. (1.2.10) for the one-dimensional error propagation, the two-dimensional version Eq. (1.2.18) is not linear in j. In fact, the leading term of $\mathbf{e}_{ic,j}$ at the center of the mesh [$ic = (I + 1)/2$] may be shown to be $[2(1 + \alpha)]^{j-2}$. (Bank and Rose [20, 23] show $\mathbf{e}_{ic,j} \cong 5.83^j$ for $\alpha = 1$.) For large J, this means that the ability of the method to resolve the error at $j = 2$ is limited by machine round-off error. For example, with $\alpha = 1$ ($\Delta x = \Delta y$) a unit error at ic for large I causes $\mathbf{F}_{ic} = 2.04 \times 10^{21}$ at $J = 30$. This means that on a computer with a 48 binary-bit characteristic, which gives about 14.5 decimal significant figures, a change of only 1 bit in $\Psi'_{ic,J}$ gives a change of approximately $(2 \times 10^{-14}) \times (2.04 \times 10^{21}) \cong 4 \times 10^{7}$ in $\Psi'_{ic,J}$. This means that the boundary condition on $\Psi_{ic,J}$ at $J = 30$ can generally be met with a tolerance of about $\pm 2 \times 10^{7}$, which is a rather poor resolution criterion!

This behavior puts a ceiling on the resolution ability of the method, even if the "inversion" of C were to be accomplished with perfect accuracy (see below). Consequently, an order-of-magnitude estimate of the practical limitation of the method can be found by numerically marching out Eqs. (1.2.18, 1.2.19) with unit errors at $j = 2$.

The error propagation has several fortunate aspects. For Poisson's equation, the largest (and, therefore, most limiting) error occurs in the center of the mesh, so that we need only consider $\mathbf{F}_{ic}$. Also, the effect of various conditions along the boundaries $i = 1$ and $i = I$ adjacent to the march have negligible effect on the center value even for I as small as 7 and all J, so we may neglect the I dimension and the adjacent boundary conditions as parameters of the error propagation. Finally, and most significantly, there is a strong effect of mesh aspect ratio $\beta = \Delta x / \Delta y$, which may be used to advantage. As stated above, the leading term in $\mathbf{F}_{ic}$ at J is $[2(1 + \alpha)]^{J-2}$, where $\alpha = \beta^{-2}$. A small β thus has an adverse effect on error propagation, while a large β has a favorable effect. Propagation, while a large β has a favorable effect. *In the limit of large β, the error propagation approaches that of the one-dimensional problem, which is merely linear in J,* as in Eq. (1.2.10).

A good estimate of the final error of this method is obtained by marching out the error propagation equation with unit errors at (ic, 2) and evaluating $P \equiv \log_{10}(\mathbf{F}_{ic})$, that is, the logarithm of the largest error that occurs in the center of the x coordinates. Then, for a particular computer with S significant decimal figures in its floating point operations, the value $(S - P)$ gives an estimate of the number of significant figures A in the answer at the end of the march. A detailed design chart for this mesh limitation on the Poisson equation discretized by the standard 5-point operator is given in Figure A-1 of Appendix A, which also gives details on other discrete analogs of the Laplacian. Unfortunately, these detailed design charts are somewhat difficult to use.

Instead of using the design charts in Appendix A, for most cases we can approximate the results concisely. With mesh aspect ratio β and the required accuracy $A = (S - P)$ as parameters, the following empirical equation predicts J_{max}, the maximum mesh size in the marching direction.

$$J_{max} \cong 4 + (CP - A)\beta \tag{1.2.38a}$$

CP is a computer parameter that depends on the word length S.

$$
\begin{aligned}
&CP = 10 && \text{for } S \cong 7.22 \ \text{(Workstation}*,\ \text{PC}**\ \text{single precision)} \\
&CP = 15 && \text{for } S \cong 13.85 \ \text{(Cray-2 single precision)} \\
&CP = 15.5 && \text{for } S \cong 14.15 \ \text{(Cray-YMP single precision)} \\
&CP = 16 && \text{for } S \cong 14.5 \ \text{(Cray-1S single precision)} \\
&CP = 18 && \text{for } S \cong 15.95 \ \text{(Workstation}*,\ \text{PC}**\ \text{double precision)} \\
&CP = 19 && \text{for } S \cong 16.86 \ \text{(microVAXII double precision)}
\end{aligned}
\tag{1.2.38b}
$$

(* SG Iris, Sun, HP, DEC α)
(* * IEEE Standard processors, includes Paragon, etc.)

This generally conservative equation predicts J_{max} to within 4% (or to $J_{max} \pm 1$ for $\beta \sim 1$) for the practical range of $3 \leq A \leq 6$ and for $1 \leq \beta \leq 11$. As an example, consider a Cray-YMP and $\beta = 5$. Then Eq. (1.2.38) indicates that we would have four-figure accuracy ($A = 4$) for J_{max} as large as 61. For $\beta = 3\frac{1}{3}$, we would be limited to about $J_{max} = 44$ for each sub-domain.

As another example, consider a Workstation or high-end PC using double precision calculations. For four-figure accuracy, Eq. (1.2.38) indicates $J_{max} = 74$ for $\beta = 5$, or $J_{max} = 50$ for $\beta = 3\frac{1}{3}$. These numbers also apply to each sub-domain solved on massively parallel computers like the Paragon that use Intel or similar IEEE standard processors.

Scientific calculations using Workstations and PC's are usually performed using double precision in any case, so there is no penalty involved. However, the Cray-2 and Cray-YMP computers suffer serious speed decreases with double precision, so it is not recommended. However, it is interesting to note that if computer manufacturers decided on using the clearly attainable precision of the double precision Cray-YMP as a standard, its ~ 28.6 significant figures would give 4 figure accuracy at $J_{max} = 148$ for $\beta = 5$, and $J_{max} = 100$ for $\beta = 3\frac{1}{3}$.

A dimensionless length of interest in determining the applicability of the method is $(J_{max} - 1)/\beta = Y/\Delta x$; that is, the number of x increments that we can march in the y direction. From Eq. (1.2.38a) for large β, we obtain

$$\frac{Y}{\Delta x} \cong CP - A \tag{1.2.39}$$

For large β, $Y/\Delta x$ becomes approximately independent of the mesh size J. For example, requiring four significant figures on a Cray-YMP with large β, we find that $\max Y/\Delta x \cong 11.5$, regardless of J.

This resolution error differs completely from a convergence criterion used in iterative methods. The largest resolution errors in marching methods appear on the single boundary at the end of the march, while the errors at interior points are much smaller. Thus, a resolution error in Ψ of 10^{-6} at the final boundary in the marching method is not directly comparable to, but is better than, an iterative change in Ψ of 10^{-6} in ADI or SOR*.

There are some other sources of error, but these are usually small. The inversion of C will contain some accumulated round-off errors; these can be reduced to a negligible level by using a Gaussian elimination routine that performs a double-precision iteration [31,50]. Also, the details of a particular problem (i.e., Ψ and ζ values) do not significantly affect the error provided that the boundary values of Ψ are reasonably scaled. Furthermore, these errors can be significantly reduced by one or two overall iterations of the entire marching method, repeating the march starting at some $j = jr$, with $\Psi'_{i,2}$ equal to the value from the previous iteration. The value of jr should be chosen as the index of a line where the values of the error $\mathbf{e}$ are several orders of magnitude larger than the initial values (D. Dietrich, personal communication). (In early work [8–10], we considered repeating the march starting from $jr = 2$, but this procedure usually improves the final error only slightly, less than the effect of reducing J by 1.)

In the Generalized Sweepout Method (GSM) [14], the final error vector is defined in terms of residuals. The marching equation is used to calculate to $(J - 1)$, and the correct boundary values at J are used to evaluate the residuals R_i at $(J - 1)$ from the following equation.

$$R_i = \frac{\Psi'_{i+1,J-1} - 2\Psi'_{i,J-1} + \Psi'_{i-1,J-1}}{\Delta x^2} + \frac{\Psi_{i,J} - 2\Psi'_{i,J-1} + \Psi'_{i,J-2}}{\Delta y^2} - \zeta_{i,J-1} \tag{1.2.40}$$

The influence matrix C is generated by the homogeneous counterpart of this equation, as before. Other aspects are the same as EVP. Both methods give the same order of errors. The GSM definition gives slightly smaller errors at $\Delta x \cong \Delta y$ because the equations are marched up to only $(J - 1)$ instead of J. The EVP method gives generally smaller errors for $\Delta x \gg \Delta y$ because of less arithmetic in forming $\mathbf{F}_i$, but the distinction is of little consequence.

1.2.6 Gradient, Mixed, and Periodic Boundary Conditions

The effects on stability of non-Dirichlet boundary conditions are readily evaluated. For a Neumann (gradient) condition along some boundary,

$$\frac{\partial \Psi}{\partial n} = s \tag{1.2.41a}$$

---

* The largest magnitude convergence errors of iterative methods appear at internal points, while the specified boundary values remain intact. Also, when SOR is used with the near-optimum over-relaxation parameter, a high-frequency "hash" error of this magnitude appears at internal points. This is of some consequence in fluid dynamic calculations, since the dynamic use of ψ is in the determination of velocity components by numerical differentiation of ψ. Although the residual errors of marching methods are virtually zero at interior points, the solution for ψ can have a high frequency component. For SOR solutions, and for marching solutions near the allowable error tolerance, a single final iteration of a Liebman (Gauss-Seidel) iteration (i.e., SOR with relaxation factor = 1) will effectively remove this.

where n is the direction normal to that boundary, we consider the difference analog

$$\frac{\Psi_{b+1} - \Psi_b}{\Delta n} = s \tag{1.2.41b}$$

Here, Δn is the mesh increment along n. If the boundary is placed along mesh points, then b is a point on the boundary, $(b + 1)$ is the adjacent interior point, and Eq. (1.2.41b) is the first-order approximation to Eq. (1.2.41a). If the boundary is between mesh points, then b is at a distance $\Delta n/2$ outside the region of interest, $(b + 1)$ is the adjacent interior point, and Eq. (1.2.41b) is second-order accurate.

For example, consider the reference problem with Eq. (1.2.41) applied at the boundary B3 adjacent to the march direction. After the value of an internal point $\Psi'_{2,j}$ has been calculated by Eq. (1.2.17), the boundary value $\Psi'_{1,j}$ is set as

$$\Psi'_{1,j} = \Psi'_{2,j} - s\,\Delta x \tag{1.2.42}$$

and at max $j \equiv J$. Substituting Eq. (1.2.16) and subtracting the correct condition on Ψ, we find

$$\mathbf{e}_{1,j} = \mathbf{e}_{2,j} \tag{1.2.43}$$

Thus, *any* value of s in the gradient condition $\partial\Psi/\partial n = s$ gives the boundary condition on $\mathbf{e}$ of $\partial\mathbf{e}/\partial n = 0$, which is used on the march equation for $\mathbf{e}_{i,j}$.

Note that the point (1, 2) is *not* part of the error vector. If it were, C would be singular, since $\mathbf{e}_{1,2}$ and $\mathbf{e}_{2,2}$ are linearly related. Likewise for $(I, 2)$.

Similarly, a mixed boundary condition (sometimes called Robin's condition), gives

$$\Psi + r\frac{\partial\Psi}{\partial n} = q \tag{1.2.44}$$

$$\mathbf{e} + r\frac{\partial e}{\partial n} = 0 \tag{1.2.45}$$

In finite difference form along B3, these are

$$\Psi_{1,j} = \frac{q - (r/\Delta x)\Psi_{2,j}}{1 - r/\Delta x} \tag{1.2.46}$$

$$\mathbf{e}_{1,j} = \frac{-(r/\Delta x)\mathbf{e}_{2,j}}{1 - r/\Delta x} \tag{1.2.47}$$

[The bad condition $r/\Delta x \to 1$ indicates degeneracy of Eq. (1.2.44) to $\Psi(2,j) = q$.] Similar formulations apply to an "angle" condition used in linearized potential flow and in magnetohydrodynamics, $\partial\Psi/\partial x = r\partial\Psi/\partial y$, with r given.

For Neumann or mixed conditions along any part of the adjacent boundaries like B3 or B4, Ψ' and Ψ at internal points are marched out via Eq. (1.2.17) with boundary values set from Eq. (1.2.42) or Eq. (1.2.46). The influence coefficient matrix C is generated by marching out Eq. (1.2.18) with boundary values set from Eq. (1.2.43) or Eq. (1.2.47). The effect of these conditions at B3 or B4 on the field-limiting magnitude of $\mathbf{F}_{\max}$ is negligible for reasonable I.

For gradient boundary conditions at boundary B2, at the end of the march direction, no change is required in the $\mathbf{e}$ or Ψ equations, but $\mathbf{F}$ is now defined in terms of the desired quantity. For the gradient condition Eq. (1.2.41), $\mathbf{F}_l$ is no longer equal to $\mathbf{e}_{i,j}$, but

$$\mathbf{F}_l = \frac{\mathbf{e}_{i,J} - \mathbf{e}_{i,J-1}}{\Delta y} \qquad \text{where } l = i - 1 \tag{1.2.48}$$

Then Eq. (1.2.24) is replaced by

$$\mathbf{e}_{m+1,2} = \frac{C_{ml}^{-1}\left(\mathbf{e}_{l+1,J} - \mathbf{e}_{l+1,J-1}\right)}{\Delta y} \tag{1.2.49}$$

Since this process involves taking differences of possibly large numbers, the correct value of $\Psi_{i,JL}$ is not calculated as accurately as for a Dirichlet problem, but this is not a particular difficulty (as erroneously described in [9]) if $\partial\Psi/\partial y$ is the variable of interest at J.

For gradient or mixed conditions at B1 (the beginning of the march), the matrix C is generated by first setting $\mathbf{e}_{i,2} = 1$ for some i, as before. Then $\mathbf{e}_{i,1}$ is set in accordance with Eq. (1.2.43) or Eq. (1.2.47) and $\mathbf{e}$ is marched out by Eq. (1.2.18) as before. A similar procedure is followed in the Ψ' and Ψ marches. The Neumann condition at B1 has a small beneficial effect on the error propagation. The leading term of $\mathbf{F}_{ic}$ is reduced by the factor $(2 + \alpha)/(2 + 2\alpha)$, which ranges from 1 to 1/2 as α ranges from 0 to ∞. The mixed condition has an intermediate effect, depending on the value of p.

Periodic boundary conditions in x are also accommodated easily. For concreteness, we choose the mesh so that the period runs from $i = 2$ to $i = I$. The interior points from $2 \le i \le (I-1)$ are marched one line as before, to $(j+1)$. Then we set $\Psi_{I,j+1} = \Psi_{2,j+1}$ and $\Psi_{1,j+1} = \Psi_{I-1,j+1}$ and continue to the next line of the march. Periodicity causes some large terms in C near the ends, so the likelihood of ill-conditioning of C increases. However, there are other advantages. For constant coefficient equations (more generally, for coefficients that are functions of y only), C becomes symmetric and constant along all diagonals. Thus, any row of C may be obtained by cyclic permutation of a known row. The generation of C is then performed by just one march of the error propagation (1.2.18), as contrasted to the L marches required for other boundary conditions as in Eq. (1.2.26). (For $L = M$, about 1/4 of these are null calculations that can be avoided.)

This greatly reduces the operation count for initialization; the only $O(M^3)$ contribution that remains comes from the Gaussian elimination with symmetry, giving

$$\theta_{\text{per},a(y)} \cong \left[\frac{1}{6}L^3\right] *, \left[\frac{1}{6}L^3\right] \pm \tag{1.2.50}$$

The storage for the LU decomposition of C is $L^2/2$. Periodic conditions in y, along the march direction, may be treated also, but with some difficulty (**E** and **F** are twice as long), and this should be avoided if possible.

It is perhaps worth emphasizing that it is not necessary to have the same type of boundary condition along an entire boundary. Reevaluation and subsequent inversion of C are not required unless either the mesh changes or the *type* of boundary condition changes, or the value of r changes in Robin's condition or in the angle condition (see Eq. 1.2.44 ff). Thus, the present formulation of the marching solution involves no penalty in the operation count for various combinations of the boundary conditions. The types of conditions could change from point to point without affecting the algorithm; the changes in programming are trivial, although we have had difficulty designing general software for these problems. This is in contrast to other direct methods that suffer a significant penalty in such cases. Of course, the method is not applicable without iteration to *nonlinear* boundary conditions any more than it is to nonlinear interior equations without iteration (see Chapter 8).

In the GSM method as described by Hirota et al. [14], Neumann conditions at J are set by using the specified gradient to project to a value $\Psi_{i,J+1}$ and evaluating the residual R_i as in Eq. (1.2.40) at J. This means that the solution depends on ζ at the boundary J, which is not the case with the EVP description above, with the usual iterative methods, or indeed with the *continuum* equations. However, the GSM residual definition is readily adapted to the equations used above.

Gradient boundary conditions can also be represented by 3-point differences. When used at the beginning or end of the march, their use would require some additional algebra (partial Gaussian elimination) at those rows adjacent to the boundaries ($J = 2$ and $J = JL - 1$ in the reference 2D problem). When used at boundaries adjacent to the march direction ($I = 1$ and $I = IL$ in the reference 2D problem), the modification is completely straightforward for the 5 point operator. The use with 9 point operators requires some additional discussion, and will be covered in Section 1.3.8.

1.2.7 Irregular Mesh and Variable Coefficient Poisson Equations

The marching method is easily applied to finite difference stencils in irregular mesh systems (e.g., see Salvodori and Baron [32], p. 180, for an uneven mesh stencil for $\partial^2\Psi/\partial y^2$), provided that the stencil at (i, j) being marched in the positive j direction does not use values of Ψ beyond $(j + 1)$. (However, the use of such uneven stencils is not recommended, generally.) Likewise, non-Cartesian coordinate descriptions, variable property models, and even non-Dirichlet interior boundaries, such as the junction of two dissimilar thermal conductors, are easily treated. Each of these results in a five-point finite-difference equation that can easily be solved for $\Psi_{i,j+1}$ as in Eq. (1.2.17). The error propagation characteristics obviously must be calculated separately for each case, but this is an easy task. One merely programs the error equation analogous to Eq. (1.2.18) for the mesh and equation used, and calculates $P = \log_{10} (\mathbf{F}_{ic})$ as a function of J. This can be done without the investment of time in an actual solution, i.e., setting boundary conditions, establishing and solving C, etc. For example, the cylindrical Poisson equation has slightly improved error propagation (smaller P) compared with the Cartesian. Other discrete analogs of the Laplacian are considered in Appendix A, based on [9].

1.2.8 Irregular Geometries

One of the advantages of a marching method over other direct methods is in its simple adaptation to irregular boundary geometries and varied combinations of boundary conditions. The only explanation required for the adaptation to irregular geometries is in the definition of the initial and final error vectors **E** and **F**. An example is given in Fig. 1.2.3. The indexing on **E** and **F** is not unique, but the arrangement with regard to the pairs $\mathbf{E}_p$, $\mathbf{F}_p$ for $p = 8, 9, 10, 11$ will cause the largest terms in C to cluster near the diagonal. This will improve the numerical solution for Eq. (1.2.23) and aid in the banding approximations to be described in Chapter 4.

The error characteristics would have to be worked out for each case, but since the presence of boundaries more than four or five cells from an interior march path has a relatively slight favorable effect on **e**, it may be expected that the method will frequently be limited by the longest march path of the problem. Note that partial cell treatment of irregular boundaries (see Salvadori and Baron [32]) is easily accommodated in the march equation (1.2.17). (It is possible to arrange the computation to "turn corners" during the march, but this is much more complicated to program than the approach shown in Fig. 1.2.3, and it has an adverse effect on error propagation since the march path length is extended.)

1.3 Other Second-Order Elliptic Equations

Up to this point, we have considered only the Poisson equation. We now consider the application of marching methods to other second-order elliptic equations.

1.3.1 Advection-Diffusion Equations

We have already described how variable coefficient Poisson equations can be solved by a marching method with only slight additional algebra in solving for $\Psi_{i,j+1}$ at each point in the march. Similarly, we can readily treat first derivative terms (such as advection or convection terms in fluid dynamics and heat transfer problems), but in this case some interesting qualitative features emerge.

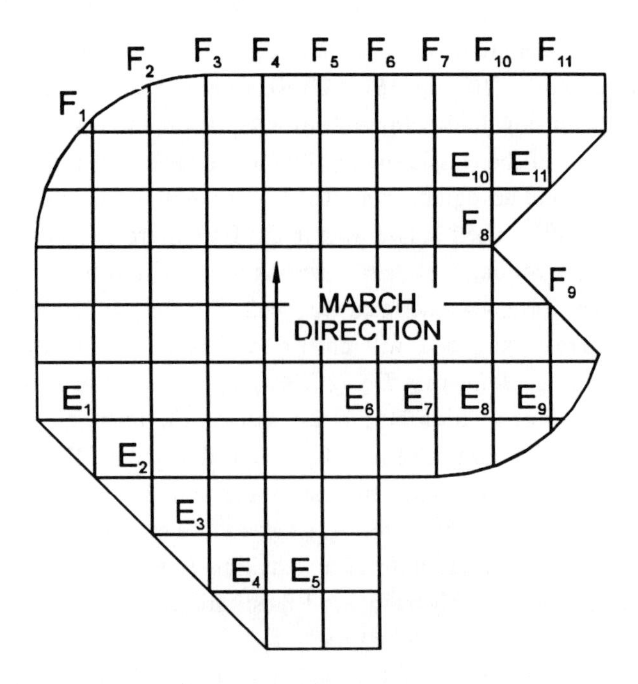

Figure 1.2.3. Marching Method for irregular geometries. E_m = components of the initial error vector; F_l = components of the final error vector.

Consider the model advection-diffusion equation

$$\mu \frac{\partial^2 \zeta}{\partial x^2} + \mu \frac{\partial^2 \zeta}{\partial y^2} - u(x,y) \frac{\partial \zeta}{\partial x} - v(x,y) \frac{\partial \zeta}{\partial y} = 0 \tag{1.3.1a}$$

In this equation, ζ could be vorticity (or a momentum component) with $\mu = 1/\mathrm{Re}$, or ζ could be temperature with $\mu = 1/\mathrm{Pe}$, etc. (The conservation form, involving terms like $\partial(u\zeta)/\partial x$, could just as easily be treated, but the nonconservation form suffices for our purposes here.) For a decoupled energy equation, u and v could be obtained previously by a steady-state fluid solution, and the task can be to obtain the steady-state temperature field ζ from Eq. (1.3.1a). Or, ζ may be vorticity, with $u = u(\zeta)$ lagged in semidirect iterations (see Chapter 8) for a steady-state solution. Using $O(\Delta^2)$ centered differences as before, and inverting the stencil to solve for $\zeta_{i,j+1}$, we obtain

$$\zeta_{i,j+1} = \frac{2(1+\alpha)\zeta_{i,j} - \alpha(1 - u_{i,j}\Delta x/2\mu)\zeta_{1+1,j} - \alpha(1 + u_{i,j}\Delta x/2\mu)\zeta_{i-1,j}}{1 - v_{i,j}\Delta y/2\mu} - \frac{(1 + v_{i,j}\Delta y/2\mu)\zeta_{i,j-1}}{1 - v_{i,j}\Delta y/2\mu} \tag{1.3.1b}$$

The groups $u\Delta x/\mu \equiv \mathrm{Re}_{cx}$ and $v\Delta y/\mu \equiv \mathrm{Re}_{cy}$ are the well-known and ubiquitous cell Reynolds numbers [10], or cell Peclet numbers, in the x and y directions, respectively. It is recognized that the condition of $Re_c = 2$ has special significance to numerics. For example, in previous work ([10], pp. 161–165) we showed that, in a centered difference analog of the one-dimensional advection-diffusion equation, the condition $Re_c = 2$ is the finite difference analog of the classic singularity (i.e., reduction of order of the equation) that occurs in the continuum equation as $Re \rightarrow \infty$ ($\nu \rightarrow 0$). This special cell Reynolds number condition manifests itself in two ways in the EVP marching solution. First, in the marching y direction, a condition $Re_{cy} = 2$ results in a division by zero in Eq. (1.3.1b). This occurs *locally*, i.e., for any $v_{i,j}$ such that $Re_{cy} = 2$. For $Re_{cy} > 2$, nothing bad happens. Practically, the condition $Re_{cy} \cong 2$ will destroy accuracy, but for $Re_{cy} < 2$ the error propagation is affected so little that Eq. (1.2.38a) applies. Second, in the x direction normal to the march, the condition $Re_{cx} = 2$ results in the term $\zeta_{i+1,j}$ being eliminated from the stencil, so there is no upstream influence at that point. For a model equation with $u_{i,j}$ constant (this is of possible interest in the split NOS semidirect method in Chapter 8) and $Re_{cx} = 2$ everywhere, the result is an influence coefficient matrix C that is *upper triangular*, i.e., the finite-difference equation is parabolic in x. This matrix equation can be solved immediately by forward substitutions, i.e., it is a marching problem *in* x.

Both of these manifestations of $Re_{cx} = 2$ and $Re_{cy} = 2$ are consistent with the interpretation that they are the finite-difference equivalents of $Re = \infty$ in the continuum equations. Going *beyond* these conditions of $Re_{cx} = 2$ or $Re_{cy} = 2$ causes no aberrations in the solutions, provided that the flow structure is such that diffusion is not important in that direction. However, in that case we must ask whether a simplified set of equations (e.g., boundary-layer equations or parabolic-marching equations) might be more appropriate.

In general, the effect on the error propagation of first-derivative advection terms is complicated, involving cell Reynolds number interpretations and proper scaling of solutions for large Reynolds numbers (e.g., see Roache [33] and Chapter 8). For $Re_c < 2$, however, the error propagation is slightly improved from that of the Poisson equation.

1.3.2 Upwind Differences

The use of *upwind differences* for advection terms is attractive from the viewpoint of physical reasoning, since they posses the *transport* property [10], i.e., disturbances are advected only in the flow direction. Second-order upwind differences (see [34-37] for early applications) can be accurate and are presently still under development [38]. First-order (2-point) upwind differences are notoriously inaccurate, to the point of being categorically rejected for publication in the *ASME Journal of Fluids Engineering* [39]. In spite of this, the first-order upwind difference method is still used by many authors (and is essential to most commerical CFD codes) because of its robustness. It is, therefore, something of a shock to find that upwind differences can be disastrously *destabilizing* to a marching solution.

The usual first-order upwind forms are

$$u\left.\frac{\partial\zeta}{\partial x}\right|_{ij} = u_{i,j}\frac{\zeta_{i,j} - \zeta_{i-1,j}}{\Delta x} \quad \text{for } u_{i,j} > 0 \tag{1.3.2a}$$

$$v\left.\frac{\partial\zeta}{\partial y}\right|_{ij} = v_{i,j}\frac{\zeta_{i,j} - \zeta_{i,j-1}}{\Delta y} \quad \text{for } v_{i,j} > 0 \tag{1.3.2b}$$

Using these and $O(\Delta^2)$ centered differences for diffusion terms in Eq. (1.3.1a) gives

$$\begin{aligned}\zeta_{i,j+1} &= \left[2(1+\alpha) + \frac{\alpha u_{i,j}\Delta x}{\mu} + \frac{v_{i,j}\Delta y}{\mu}\right]\zeta_{i,j} \\ &\quad - \alpha\left[\zeta_{i+1,j} + \left(1 + \frac{u_{i,j}\Delta x}{\mu}\right)\zeta_{i-1,j}\right] - \left[1 + \frac{v_{i,j}\Delta y}{\mu}\right]\zeta_{i,j-1}\end{aligned} \tag{1.3.3}$$

As in the one-dimensional problem ([10], p. 165), nothing particular happens at Re_{cx} or $\mathrm{Re}_{cy} = 2$. Significantly, there is no possibility for a division by zero. However, simple calculations show that the error growth is greatly amplified in comparison to the $O(\Delta^2)$ analog. The one-dimensional case is obtained from Eq. (1.3.3) by setting $\alpha = 0$ and $u = 0$. For $\mathrm{Re}_{cy} > 1$, the error propagation equation changes from one that is merely linear in j to one in which the leading term is $\mathbf{e}_j = \mathbf{e}_2(\mathrm{Re}_{cy})^{j-1} + \cdots$.

The destabilizing effect can be understood in terms of a comparison to the $O(\Delta^2)$ centered difference calculations, using the well-known analysis (e.g., [10], pp. 65–66) for the *artificial viscosity* of the upwind differences, which can be shown to effectively introduce so-called artificial viscosity coefficients $\nu_{ex} = u\Delta x/2$ and $\nu_{ey} = v\Delta y/2$. While it is helpful in interpreting the highly inaccurate artificial damping of the upwind differencing method, this interpretation has some difficulties, partly because of the ambiguity between transient and steady-state results [10] and partly because of the directionality of ν_{ex} and ν_{ey}. Generally, $u\Delta x \neq v\Delta y$, so the coefficients are only qualitatively like viscosity coefficients in two dimensions, since true physical (Newtonian) viscosity gives a nondirectional ν. As suggested by T. A. Porsching (personal communication), a more precise interpretation is in terms of an *artificial ellipticity*. Once said, the interpretation of the instability of marching is immediate; the upwind difference in x adds ellipticity in x, which further increases the Hadamard instability or "ill-posedness" of marching out an elliptic problem.

If we group the coefficients as in the artificial viscosity analysis and compare the upwind equation (1.3.3) with the centered difference equation (1.3.1b), we find that the upwind equation is identical to the centered difference equation with the cell aspect ratio

β decreased by a factor of $1/(1 + Re_{cx})$. Usually, upwind differences are not used unless $Re_{cx} \geq 2$. Even at this value, we have effectively reduced the cell aspect ratio by a factor of 1/3, which accounts for the destabilizing effect.

However, upwind differencing in the march direction y adds ellipticity in y (compared to that in x) thus making the error propagation approach the merely linear growth of the 1D problem. This stabilizing effect of marching into the upwind direction was noted and studied by Gartland [40,41].

1.3.3 Turbulence Terms

The addition of an eddy viscosity model for a boundary layer along x adds an additional y diffusion term T_μ to Eq. (1.3.1a), where

$$T_\mu = \frac{\partial}{\partial y} \epsilon(x,y) \frac{\partial \zeta}{\partial y} \tag{1.3.4}$$

As before, the centered difference analog is readily inverted for $\zeta_{i,j+1}$. The eddy viscosity term ϵ is a spatially varying and nonlinear term. In our semidirect iteration [33; see also Chapter 8] it is lagged, so for the present purposes it can be considered linear. The more important point here is that $\epsilon \gg \mu$ in Eq. (1.3.1a).

This added turbulence term has a beneficial effect on the error propagation because it increases the ellipticity in y compared with x. This drives the error propagation toward the one-dimensional case, which for constant ϵ gives merely linear error growth. For physically realistic models of ϵ, the improvement is so good that stability considerations can practically be ignored. This would not necessarily be the case for more elaborate turbulence models, which have yet to be analyzed, but it is consistent with the physics, regardless of the particular model. The resolution requirement in y is increased, but the added mesh points in the same physical distance (i.e., increasing J while proportionately decreasing Δy) have little effect on the final errors, as shown in Eq. (1.2.39).

1.3.4 Fibonacci Scale

The accurate calculation of skin friction and wall heat transfer coefficients in a turbulent boundary layer requires high resolution (small Δy) near the wall, as is well known. If a uniform grid at this constant Δy is used throughout the boundary layer, a prohibitively large number of grid points results. Consequently, the most efficient turbulent boundary layer calculations are performed in an expanding grid. A common prescription is

$$\Delta y_{j+1} = \tilde{K} \Delta y_j \tag{1.3.5}$$

where $\tilde{K}$ is constant. This prescription can result in a loss of formal second-order accuracy when the usual three-point finite difference equations are used. Satisfactory accuracy is achieved by making the mesh expand slowly; Harris and Morris [42], for example, used $\tilde{K} = 1.02$. If it were not for this formal loss of accuracy, a more rapidly expanding mesh would be appropriate for typical turbulent boundary layers; Keller and Cebeci [43], for example, used $\tilde{K} \cong 1.7$ in a calculation using the "box" scheme, which suffers no formal loss of accuracy for $\Delta y_2 \neq \Delta y_1$.

The following is an alternate (and quite specialized) approach to achieving the rapidly expanding mesh while retaining formal second-order accuracy and using only the customary finite-difference equations without a coordinate transformation for diffusion. It is based on a Fibonacci scale, and it introduces some novel aspects into the marching solution.

The Fibonacci sequence $\mathcal{F}_l$ is given by [44]

$$\mathcal{F} = 1,\ 1,\ 2,\ 3,\ 5,\ 8,\ 13,\ 21,\ 34, \cdots \tag{1.3.6a}$$

$$\mathcal{F}_1 = 1 \qquad \mathcal{F}_2 = 1 \qquad \mathcal{F}_l = \mathcal{F}_{l-1} + \mathcal{F}_{l-2} \tag{1.3.6b}$$

That is, each Fibonacci number is the sum of the preceding two numbers in the sequence. If we let $\Delta y_j \propto \mathcal{F}_j$, or

$$y_{j+1} - y_j \equiv \Delta y_j = \Delta y_1 \mathcal{F}_j \tag{1.3.7}$$

then we have an expanding mesh as shown in Fig.1.3.1.

We can now write the usual second-order difference equations for constant Δy between $(j + 1)$ and $(j - 2)$, not using the value at $(j - 1)$:

$$\left.\frac{\partial \zeta}{\partial y}\right|_j = \frac{\zeta_{j+1} - \zeta_{j-2}}{y_{j+1} - y_{j-2}} = \frac{\zeta_{j+1} - \zeta_{j-2}}{2\Delta y_j} \tag{1.3.8a}$$

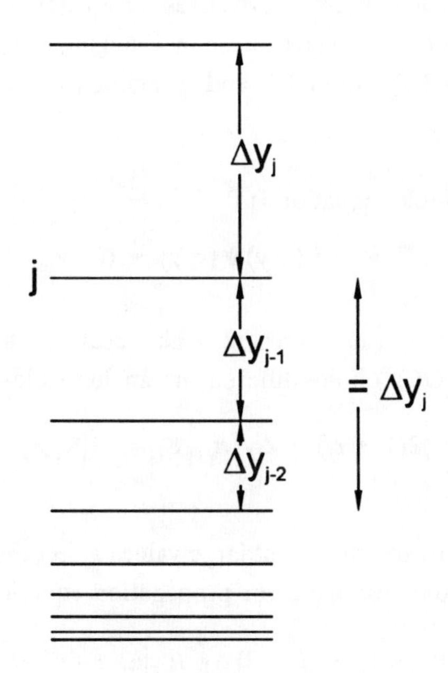

Figure 1.3.1. Expanding mesh based on the Fibonacci sequence.

$$\left.\frac{\partial^2 \zeta}{\partial y^2}\right|_j = \frac{\zeta_{j+1} - 2\zeta_j + \zeta_{j-2}}{\Delta y_j^2} \tag{1.3.8b}$$

For large j, the Fibonacci sequence gives $\tilde{K} = 1.618$ in Eq. (1.3.5). The mesh expansion can be halted at any point and the constant Δy equations used, as they are at $j = 2$.

A marching algorithm is readily adapted to this Fibonacci scale in an obvious way, by replacing indexes $(i, j - 1)$ by $(i, j - 2)$ and Δy by Δy_j in march equations like Eqs. (1.2.17), (1.3.1b), etc. (This simple prescription will also apply to convert a program with *cross derivatives* to the Fibonacci scale; see the section on cross derivatives below.) However, the stability is adversely affected because Δy_1 is so small that we again approach the Hadamard instability of the continuum equations. This is remedied by *reversing the march direction,* marching in from $\zeta_{i,J}$ to $\zeta_{i,1}$. Since Δy_{J-1} is large in comparison to Δy_1, the final value at $\zeta_{i,1}$ is not so sensitive to the initial guess at $\zeta_{i,J-1}$. The march equation corresponding to Eq. (1.3.1b) is

$$\zeta_{i,j-2} = \frac{2(1 + \alpha_j)\zeta_{i,j} - \alpha_j(1 - u_{i,j}\Delta x/2\nu)\zeta_{i+1,j} - \alpha(1 + u_{i,j}\Delta x/2\nu)\zeta_{i-1,j}}{1 + v_{i,j}\Delta y/2\nu} - \frac{(1 - v_{i,j}\Delta y/2\nu)\zeta_{i,j+1}}{1 + v_{i,j}\Delta y/2\nu} \tag{1.3.9}$$

where $\alpha_j = (\Delta y_j/\Delta x)^2$. To reach the last point on the wall at $y = 0$, the left-hand side of Eq. (1.3.9) is replaced by $\zeta_{i,j-1} \equiv \zeta_{i,1}$.

In summary, the inclusion of first derivatives, cross derivatives, and/or Helmholtz terms is accommodated in this Fibonacci mesh by changing $(i, j - 1)$ to $(i, j - 2)$, changing Δy to Δy_j given by Eqs. (1.3.6) and (1.3.7) and reversing the march direction.

1.3.5 Helmholtz Terms

The generalized Helmholtz equation is

$$\nabla^2 \Psi = h(x, y)\Psi(x, y) + \zeta(x, y) \tag{1.3.10}$$

where the coefficient $h(x, y)$ can represent a chemical reaction term or some other "source-sink" term. The $O(\Delta^2)$ finite-differencing analog yields the marching equation

$$\Psi_{i,j+1} = \Delta y^2 \zeta_{i,j} + \left[2(1 + \alpha) + \Delta y^2 h_{i,j}\right]\Psi_{i,j} - \alpha(\Psi_{i+1,j} + \Psi_{i-1,j}) - \Psi_{i,j-1} \tag{1.3.11}$$

Small values of $h_{i,j}$ are easily tolerated, but large values affect the stability adversely. For $|\Delta y^2 h_{i,j}| \ll 1$, the one-dimensional error propagation equation is

$$\mathbf{e}_j = \left[(j - 1) + (j - 2)\Delta y^2 h_{i,j}\right]\mathbf{e}_2 + O\left(\Delta y^2 h_{i,j}\right)^2 \tag{1.3.12}$$

Thus, small negative values of h_{ij} can actually help the stability. For $|\Delta y^2 h_{i,j}| \gg 1$, the leading term is $\mathbf{e}_j = 4(\Delta y^2 h_{i,j})^{j-2}\mathbf{e}_2 + \cdots$. This exponential growth in j quickly destroys accuracy, even in the one-dimensional case.

The favorable situation of $|\Delta y^2 h_{i,j}| < 1$ is relevant to the three-dimensional Poisson solutions by the EVP-FFT method discussed in Chapter 5, and in slowly time varying problems discussed in Chapter 8.

The same remarks apply to the eigenfunction problem wherein $h(x,y) = -\lambda$ is constant. The marching method may be used to calculate the eigenfunction, but is not applicable to the more difficult problem of determining the eigenvalues λ.

1.3.6 Cross Derivatives

Cross or mixed derivative terms like $\partial^2(\,)/\partial x\, \partial y$ typically arise because of tensor properties or nonorthogonal coordinate transformations, which are now frequently used to handle arbitrary boundary shapes with finite-difference methods. For example, when a channel with arbitrary differentiable contours of the upper and lower walls $\tilde{y}_u(x)$ and $\tilde{y}_l(x)$ is mapped onto a rectangular region in (x, y) by the linear transformation

$$y = \frac{\tilde{y} - \tilde{y}_1}{\tilde{y}_u - \tilde{y}_l} \tag{1.3.13}$$

then the Poisson equation (1.2.1) transforms to an equation of the form

$$\frac{\partial^2 \Psi}{\partial x^2} + P_1 \frac{\partial^2 \Psi}{\partial y^2} + P_2 \frac{\partial^2 \Psi}{\partial x\, \partial y} + P_3 \frac{\partial \Psi}{\partial y} = \zeta \tag{1.3.14}$$

where the parameters of the transformation P_1, P_2, and P_3 are functions of x [33]; more generally, they may be functions of x and y. Cross derivatives also arise in Cartesian coordinate descriptions from non-isotropic diffusion terms such as compressible viscous flows, heat conduction in anisotropic materials, electrical discharge in non-uniform plasmas, dispersion in porous media, etc.

The usual $O(\Delta^2)$ centered difference analog for the cross derivative is

$$\left.\frac{\partial^2 x}{\partial x \partial y}\right|_{ij} = \frac{\Psi_{i+1,j+1} - \Psi_{i-1,j+1} - \Psi_{i+1,j-1} + \Psi_{i-1,j-1}}{4\Delta x\, \Delta y} + O(\Delta^2) \tag{1.3.15}$$

When this term is added to the usual 5-point Poisson analog Eq. (1.2.15), a nine-point stencil results. We cannot solve a march equation explicitly for $\Psi_{i,j+1}$ in terms of previously known (marched) values at j and $(j-1)$. Rather, we can solve for the triple $\Psi_{i-1,j+1}$, $\Psi_{i,j+1}$ and $\Psi_{i+1,j+1}$. From Eq. (1.3.15) we obtain

$$\frac{P_2}{4\beta P_1}\Psi_{i+1,j+1} + \left[1 + \frac{P_3}{P_1}\frac{\Delta y}{2}\right]\Psi_{i,j+1} - \frac{P_2}{4\beta P_1}\Psi_{i-1,j+1} = \frac{\Delta y^2}{P_1}\zeta_{i,j} + \left[2 + 2\frac{\alpha}{P_1}\right]\Psi_{i,j}$$
$$- \frac{\alpha}{P_1}\left(\Psi_{i+1,j} + \Psi_{i-1,j}\right) - \left[1 - \frac{P_3}{P_1}\frac{\Delta y}{2}\right]\Psi_{i,j-1} + \frac{P_2}{4\beta P_1}\left(\Psi_{i+1,j-1} - \Psi_{i-1,j-1}\right) \quad (1.3.16)$$

where $\alpha = (\Delta y/\Delta x)^2$ and $\beta = \Delta x/\Delta y$, as before. The left-hand side may be solved implicitly for the entire line of L new values at $(j + 1)$ using a tridiagonal algorithm. Thus, the march proceeds similar to a line SOR solution (e.g., see [10]).
The necessity for the tridiagonal solution adds to the operation count, aside from the requirement for extra arithmetic to evaluate $\partial^2\Psi/\partial x\,\partial y$. A tridiagonal solution for L equations has the following operation count ([27], p. 167):

$$\theta_6 = 3(L - 1) *,\ 3(L - 1) \pm,\ 2(L - 1) \div \quad (1.3.17)$$

This applies for each of M lines in a march. Assuming that the coefficients of Eq. (1.3.16) are prestored, the homogeneous error propagation equation obtained from it requires 3 * and 6±. (Unlike the march of the explicit five-point equation without cross derivatives, there are no null calculations in this implicit march.) Thus, the L marches of this equation to establish C give an operation count of

$$\theta_7 = L\left(M\theta_6 + LM[3*, 6\pm]\right) \quad (1.3.18)$$

Neglecting $1 \ll L$ in Eq. (1.3.17) and substituting in Eq. (1.3.18) gives

$$\theta_8 = 6L^2M*,\ 9L^2M\pm,\ 2L^2M\div \quad (1.3.19)$$

The operation count for the LU decomposition is unchanged from θ_3 given by Eq. (1.2.27). The total operation count required to initialize the solution to Eq. (1.3.16) with cross derivatives is given by θ_8 from Eq. (1.3.19) and θ_3 from Eq. (1.2.27) and is given by the following equation.

$$\theta_{\times,\mathrm{init}} = \left[\frac{1}{3}L^3 + 6L^2M + \frac{1}{2}L^2\right]*,\ \left[\frac{1}{3}L^3 + 9L^2M + \frac{1}{2}L^2\right]\pm,$$
$$\left[2L^2M + \frac{1}{2}L^2\right]\div \quad (1.3.20)$$

For repeat solutions, we have two marches of Eq. (1.3.16); the right-hand side requires 4 * and 6± at each of $L \times M$ points, and M lines of θ_6 from Eq. (1.3.17), giving

$$\theta_9 = 2\left(M\theta_6 + LM[4*, 6\pm]\right) \quad (1.3.21)$$

Neglecting $1 \ll L$ in Eq. (1.3.17) and substituting in Eq. (1.3.21) gives

$$\theta_{10} = 14LM*,\ 18LM\pm,\ 4LM\div \tag{1.3.22}$$

The solution $\mathbf{e}_{i,2}$ from the initialization LU decomposition is unchanged from Eq. (1.2.33). Thus, the repeat solutions require total operations of θ_{10} from Eq. (1.3.22) and θ_5 from Eq. (1.2.33) plus some lower-order terms that we neglect, giving

$$\theta_{\times,\mathrm{rep}} = \left(14LM + L^2\right)*,\ \left(18LM + L^2\right)\pm,\ 4LM\div \tag{1.3.23}$$

For $L = M$, and counting each $*$, $\pm$, or $\div$ as on operation, we obtain from Eqs. (1.3.20) and (1.3.23) the total operations for the marching solution of equations like Eq. (1.3.16) with cross derivatives.

$$\theta_{\times,\mathrm{init}} = 17\frac{2}{3}M^3 + \frac{3}{2}M^2 \tag{1.3.24a}$$

$$\theta_{\times,\mathrm{rep}} = 38M^2 \tag{1.3.24b}$$

Thus, the operation count for repeat solutions for cross derivative equations with $L = M$ are still optimal, i.e., proportional to the number of mesh points. Comparing Eq. (1.3.24) above for cross derivatives with Eq. (1.2.48) for the simple Poisson equation shows a penalty ratio of about 4.4 for the initial solution and about 2.7 for repeat solutions.

A skewed seven-point stencil for $\partial^2(\ )/\partial x\,\partial y$ [44,45] can be used instead of Eq. (1.3.13), resulting in an upper (lower) bidiagonal matrix for Ψ_{j+1} that can be solved by back (forward) substitution starting from the right (left) boundary value. However, the form used here is a more easily justifiable form to use for general equations.

1.3.7 Gradient Boundary Conditions and Cross Derivatives

The use of gradient boundary conditions when cross derivatives are present requires two special considerations.

The first special consideration is for the 3-point normal derivative. In Section 1.2.6, we noted how the 3-point discretization along boundaries adjacent to the march direction ($I = 1$ and $I = IL$ in the reference 2D problem) is easily accommodated for the 5-point operator. However, for the 9-point operator (resulting from cross-derivatives), this would increase the band width of the marching procedure, so that the line SOR march would require a pentadiagonal solver instead of a tridiagonal solver. In [33,46] this increase in computer time was avoided by using special Gaussian elimination at the near-boundary points. This treatment was satisfactory for the transformed Poisson equation, but for the vorticity equation at high Reynolds number (in which the corner terms of the 9-point matrix become very small) a deteriorating condition of the matrix resulted in large round-off errors in the finest (81 × 81) mesh. The problem was solved by using only 2-point $O(\Delta)$ gradient conditions in the direct marching algorithm, with the 3-point correction to $O(\Delta^2)$ accuracy being lagged in the nonlinear iterations. This procedure converged very quickly and is more robust than the direct 3-point method. Also, the matrix description uses only parts of the same stencil [Eq. (1.2.15)] used at interior points, whereas the 3-point normal gradient would require special matrix description at boundaries. For these reasons, only the 2-point gradient formulation is allowed in the GEM codes (see Chapter 7).

The second special consideration arises specifically in the use of gradient boundary conditions in nonorthogonal grids, which give rise to cross derivatives at interior points. The requirement for marching in (say) the J direction is that the boundary conditions must be separable in J. Using part of a notation that will be used in Chapter 6, we can write the boundary stencil at $I = 1$ as

$$\begin{bmatrix} & \vdots & \\ & (C2) & \\ & C5 & C6 \\ & (C8) & \\ & \vdots & \end{bmatrix} \mathbf{F}_{ij} = C10$$

The general form of this equation allows for all linear combinations of boundary conditions such as Dirichlet (specified function value **F**), Neumann (specified normal gradient of **F**), ratio of mixed derivatives, etc. However, the requirement for separability of boundary conditions in the assumed marching direction y indicates that C2 cannot be used at the side boundaries for a march in increasing J, nor can C8 be used for a march in decreasing J. At first glance, this limitation appears only to restrict the evaluation of the tangential derivatives at the side boundaries to a first-order 2-point formulation. However, a more serious ramification occurs that has practical importance for solutions in nonorthogonal boundary fitted coordinate problems [33,46,47].

Consider a Neumann boundary condition of $\mathbf{F}_n = 0$ at $I = 1$, where n is the direction normal to the left boundary. For a Cartesian or other orthogonal grid, this condition would involve only derivatives in the I direction, so C2 and C8 would both be zero. However, for a nonorthogonal grid at $I = 1$, the normal derivative condition transforms to

$$\frac{dF}{dn} = \left(a\,\mathbf{F}_X - b\,\mathbf{F}_Y\right) / V\sqrt{a}$$

where Y is the tangential coordinate direction (along J), X is the quasi-normal coordinate direction (along I), a and b are coefficients of the transformation from the Cartesian to the general nonorthogonal grid, and V is the Jacobian of the transformation. For a locally orthogonal grid, $b = 0$. For $b \neq 0$, the requirement for either $C2 = 0$ or $C8 = 0$ is set by the choice of the march direction, which may be dictated by stability and resolution requirements. (For example, an expanding grid designed for high resolution of a boundary layer at $J = 1$ makes a march in decreasing J direction desirable from stability considerations; see Section 1.3.3). However, depending on the curvature of the left boundary (the sign of b at $I = 1$), this can be *analogous to downwind* differencing along the left boundary, and can produce *oscillations* in the solution for **F**.

In analogy with the well-known phenomena in computational fluid dynamics [10], we would anticipate that this behavior could arise even using centered differences for $\mathbf{F}_Y$ at large values of $|b|$. The difficulty appears to be inherent to nonorthogonal grids, not just to marching solution methods, but the peculiar requirement of the marching methods for directional differencing at boundaries adjacent to the march direction exaggerates the problem.

The "cure" is to have a nearly orthogonal grid near boundaries, giving $b \cong 0$. For a symmetry boundary, this is of course a reasonable condition. Note, however, that simply setting $b \cong 0$ by reflection can give a discontinuity in the grid that will slow truncation error convergence. However, consider boundary fitted coordinates in a highly skewed rhomboid geometry, and a heat conduction problem with an adiabatic wall. The objective of locally orthogonal coordinate lines will be incompatible with the requirement for boundary-fitted coordinates, at least near the corners, and the adiabatic (zero-gradient) condition will cause difficulty, especially for the marching methods.

The second order accurate solution, involving both coefficients C2 and C8 at boundaries, can be obtained by deferred corrections, lagging the difference between the one-sided and centered forms for $\mathbf{F}_Y$. It is even more robust, for geometries in which b may change sign along the boundary, to lag the entire tangential derivative $\mathbf{F}_Y$, along with deferred corrections for the 3-point form of $\mathbf{F}_X$ and any nonlinearities, and this is now our standard procedure. Note, however, that the marching method now cannot be considered a direct method for gradient boundary conditions adjacent to the march direction in a nonorthogonal grid for certain combinations of boundary curvature and marching direction.

1.3.8 Interior Flux Boundaries

When a diffusion equation is solved over a region with discontinuous material properties, an interior flux boundary condition arises. For example, consider the region shown in Figure 1.3.3, in which the dotted line represents the interior boundary where two different materials meet. In each separate region, the elliptic problem is again given by the Poisson equation,

$$\nabla^2 \Psi = \zeta \tag{1.3.25}$$

where Ψ might be temperature, electric potential, etc. But along the interface, where the diffusion coefficient (heat conductivity, electrical conductivity, etc.) changes, this equation is replaced by some interface condition.

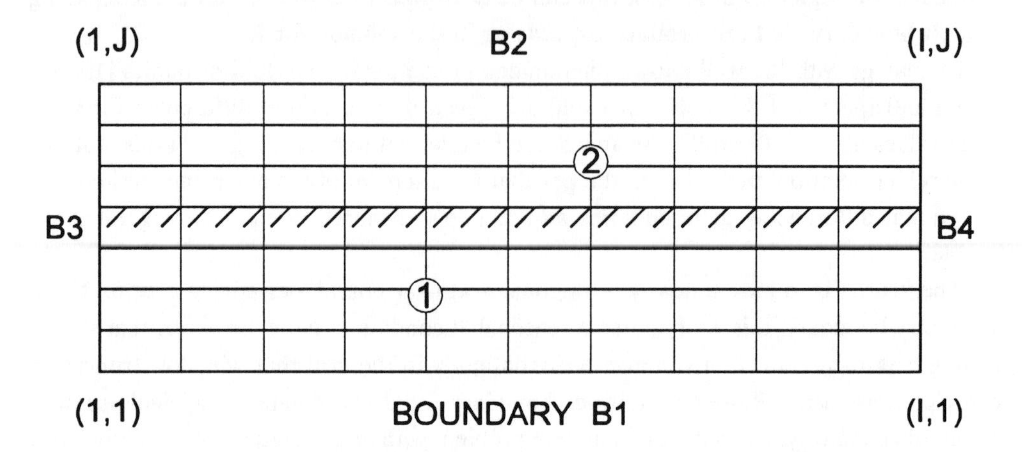

Figure 1.3.3. Interior Boundary Flux condition between regions ① *and* ②.

This situation is readily handled by the marching methods. The interface condition is usually just a special case of the more general problem

$$\nabla \cdot \sigma \nabla \Psi = \zeta \tag{1.3.26}$$

where σ changes in step fashion at the interface. The finite difference analog of this equation is readily constructed and marched out. Likewise, jump conditions such as

$$\Psi_{①} - \Psi_{②} = K\Psi_{①} \tag{1.3.27}$$

are easily treated, at no degradation in stability.

Note, however, that a true boundary *condition* cannot be marched through, as previously mentioned for 1D in Section 1.2.1. Supposing that the interior boundary in Figure 1.3.3 actually had specified values of Ψ or $\partial\Psi/\partial n$ on it. (For example, it could be a specified voltage in an electrostatics problem.) Clearly, we now have two separate boundary value problems, one in region ①, and the other in region ②. For convenience, it might be solved as one problem using a point iterative method like SOR, with the governing elliptic equation replaced by

$$\Psi_{i,j} = \Psi_{IB} \tag{1.3.28}$$

along the interface, with the loop indexes running over the combined region. In SOR, the convergence rate is improved by such interior boundaries, since each component problem is smaller. But any attempt to march through the interior boundary will give $C_{lm} = 0$, since the error information cannot propagate through the interior boundary, giving an indeterminate problem.

This limitation occurs in a practical problem when we actually have nonspecified interior boundaries in a nonlinear problem, but specified interior values within a nested linearization, using semidirect methods (Chapter 8). This has occurred in using semidirect/marching methods to solve adaptive grid generation equations along an interior boundary [48].

1.4 Closing Note for Chapter 1

The material in this Chapter 1 was originally presented in Roache [49].

References for Chapter 1

1. R. W. Hockney, A Fast Direct Solution of Poisson's Equation Using Fourier Analysis, *Journal of the Association Computing Machinery*, Vol. 12, 1965, pp. 95-113.
2. R. W. Hockney, The Potential Calculation and Some Applications, in B. Alder, S. Fernbach, and M. Rotenberg (eds.), *Methods in Computational Physics, Vol. 9, Plasma Physics*, Academic Press, New York, 1970, pp. 134-210.
3. O. Buneman, A Compact Non-Iterative Poisson Solver, SUIPR Report 294, Stanford University, Stanford, CA, May 1969.
4. B. L. Buzbee, G. H. Golub, and C. W. Nielson, The Method of Odd-Even Reduction and Factorization with Application to Poisson's Equation, Los Alamos Science Laboratory, Report LA-4141, Los Alamos, NM, 1969.
5. F. W. Dorr, The Direct Solution of the Discrete Poisson Equation on a Rectangle, *SIAM Review*, Vol. 12, No. 2, Apr. 1970, pp. 248-263.
6. R. C. LeBail, Use of Fast Fourier Transforms for Solving Partial Differential Equations in Physics, *Journal of Computational Physics*, Vol.9, No. 3, June 1972, pp. 440-465.
7 H. F. Weinberger, *Partial Differential Equations,* Blaisdell Publishing Co., New York, 1965, p. 51.
8. P. J. Roache, A New Direct Method for the Discretized Poisson Equation, in M. Holt (ed.), *Lecture Notes in Physics, Vol. 8, Proc. Second International Conference on Numerical Methods in Fluid Mechanics*, Springer, New York, 1971, pp. 48-53.
9. P. J. Roache, A Direct Method for the Discretized Poisson Equation, Sandia National Laboratory, Report SC-RR-70-579, Albuquerque, NM, 1971.
10. P. J. Roache, *Computational Fluid Dynamics*, rev. printing, Hermosa Publishers, Albuquerque, NM, 1976.
11. H. Ishizaki, *On the Numerical Solution of Harmonic, Biharmonic and Similar Equations by the Difference Method Not Through Successive Approximations*, Disaster Prevention Research Institute Bulletin 18, Kyoto University, Kyoto, Japan, Aug. 1957.
12. W. G. Bickley and J. McNamee, Matrix and Other Direct Methods for the Solution of Systems of Linear Difference Equations, *Proc. Royal Society of London, Ser. A*, Vol. 252, Jan. 1960, pp. 69-131.

13. M. L. Booy, A Noniterative Numerical Solution of Poisson's and Laplace's Equations with Applications to Slow Viscous Flow, *Journal of Basic Engineering*, Dec. 1966, pp. 725-733.
14. I. Hirota, T. Tokioka, and M. Nishiguchi, A Direct Solution Of Poisson's Equation by Generalized Sweep-out Method, *Journal of the Meteorological Society of Japan*, Ser. II, Vol. 48, No. 2, Apr. 1970, pp. 161-167.
15. S. Schechter, Quasi-Tridiagonal Matrices and Type-Insensitive Difference Equations, *Quarterly Applications in Mathematics*, Vol. 19, 1960, pp. 285-295.
16. J. W. Lucey and K. F. Hansen, A Stable Method of Matrix Factorization, *Transactions American Nuclear Society*, Vol. 7, 1964, p. 259.
17. R. Coleman, The Numerical Solution of Linear Elliptic Equations, *Journal of Lubrication Technology*, Ser. F, Vol. 90, 1968, pp. 773-776.
18. E. Angel and R. Bellman, *Dynamic Programming and Partial Differential Equations*, Academic Press, New York, 1972.
19. F. W. Dorr, The Direct Solution of the Discrete Poisson Equation in $O(N^2)$ Operations, *SIAM Review*, Vol. 17, No. 3, 1975, pp. 412-415.
20. R. E. Bank and D. J. Rose, An $O(n^2)$ Method for Solving Constant Coefficient Boundary Value Problems in Two Dimensions, *SIAM Journal of Numerical Analysis*, Vol. 12, No. 4, 1975, pp. 529-539.
21. D. Dietrich, B. E. McDonald, and A. Warn-Varnas, Optimized Block Implicit Relaxation, *Journal of Computational Physics*, Vol. 18, No. 4, 1975, pp. 421-439.
22. S. Y. K. Yee, An Efficient Method for a Finite-Difference Solution of the Poisson Equation on the Surface of a Sphere, *Journal of Computational Physics*, Vol. 22, No. 2, 1976, pp. 215-228.
23. R. E. Bank and D. J. Rose, Marching Algorithms for Elliptic Boundary Value Problems. I. The Constant Coefficient Case, Aiken Computation Laboratory, TR 14-75, Harvard University, Cambridge, MA, 1975.
24. R. E. Bank, Marching Algorithms for Elliptic Boundary Value Problems. II. The Non-Constant Coefficient Case, Aiken Computation Laboratory, TR 16-75, Harvard University, Cambridge, MA, 1975.
25. R. E. Bank, Marching Algorithms and Block Gaussian Elimination, *Proc. Argonne Conference on Sparse Matrix Computations*, Argonne, IL, 1976.
26. E. K. Blum, *Numerical Analysis and Computation: Theory and Practice*, Addison-Wesley Publishing Co., Reading, MA, 1972, p. 109.
27. G. Dahlquist and Å. Björk, *Numerical Methods in Engineering*, Translation by N. Anderson, Prentice-Hall, Inc., Englewood Cliffs, NJ, 1961.
28. G. E. Forsythe, M. A. Malcolm, and C. B. Moler, *Computer Methods for Mathematical Computations*, Prentice-Hall, Englewood Cliffs, NJ, 1977, p. 56.
29. R. A. Sweet, A Cyclic Reduction Algorithm for Solving Block Tridiagonal Systems of Arbitrary Dimension, *SIAM Journal of Numerical Analysis*, Vol. 14, No. 4, 1977, pp. 706-720.
30. W. Proskurowski and O. Widlund, On the Numerical Solution of Helmholtz's Equation by the Capacitance Matrix Method, *Mathematics of Computation*, Vol. 30, No. 135, 1976, pp. 433-468.

31. C. B. Bailey, A Guide to the Sandia Mathematical Program Library, Sandia National Laboratories Report, SC-M-69-337, Albuquerque, NM, Jan. 1970.
32. M. G. Salvadori and M. L. Baron, *Numerical Methods in Engineering*, 2nd ed., Prentice-Hall, Englewood Cliffs, NJ, 1961.
33. D. A. Anderson, J. C. Tannehill, and R. H. Pletcher, *Computational Fluid Mechanics and Heat Transfer*, Hemisphere Publishing Corporation, Washington, 1984.
34. H. S. Price, R. S. Varga, and J. E. Warren, Applications of Oscillation Matrices to Diffusion-Convection Equations, *Journal of Mathematical Physics*, Vol. 45, 1966, pp. 301-313.
35. R. F. Warming and R. M. Beam, Upwind Second Order Difference Schemes and Applications in Aerodynamic Flows, *AIAA Journal*, Vol.14, No. 9, Sept. 1976, pp. 1241-1249.
36. M. Atais, M. Wolfshtein, and M. Israeli, Efficiency of Navier-Stokes Solvers, *AIAA Journal*, Vol. 15, No. 2, Feb. 1977, pp. 263-266.
37. B. P. Leonard, A Stable and Accurate Convective Modeling Procedure based on Quadratic Upstream Differencing, *Computer Methods in Applied Mechanics and Engineering*, Vol. 19, 1979, pp. 59-98.
38. B. P. Leonard, The ULTIMATE conservative difference scheme applied to unsteady one-dimensional advection, *Computer Methods in Applied Mechanics and Engineering*, Vol. 88, 1991, pp. 17-74.
39. P. J. Roache, K. Ghia, and F. White, Editorial Policy Statement on the Control of Numerical Accuracy, *ASME Journal of Fluids Engineering*, Vol. 108, No. 1, March 1986, p. 2.
40. E. C. Gartland, Jr., Discrete Weighted Mean Approximation of a Model Convection-Diffusion Equation, *SIAM Journal of Scientific and Statistical Computing*, Vol. 3, 1982, pp. 460-472.
41. E. C. Gartland, Jr., Approximation Methods for a Nonsymmetric Singularly Perturbed Problem, *Elliptic Problem Solvers II*, G. Birkhoff and A. L. Schoenstadt, (eds.), Academic Press, New York, 1984, pp. 545-556.
42. J. E. Harris and D. J. Morris, Solution of Three-Dimensional Boundary-Layer Equations with Comparisons to Experimental Data, in R. D. Richtmyer (ed.), *Lecture Notes in Physics, Vol. 35, Proc. Fourth International Conference on Numerical Methods in Fluid Dynamics*, Springer, New York, 1975, pp. 204-211.
43. H. B. Keller and T. Cebeci, Accurate Numerical Methods for Boundary Layer Flows. II. Two-Dimensional Turbulent Flows, *AIAA Journal*, Vol. 10, No. 9, Sept. 1972, pp. 1193-1199.
44. M. Abramovich and I. A. Stegun (eds.), *Handbook of Mathematical Functions*, 9th Printing, Equation 25.3.27, Dover Publications, New York, 1970, p. 884.
45. F. B. Hildebrand, *Finite-Difference Equations and Simulations*, Prentice-Hall, Englewood Cliffs, NJ, 1968, p. 95.
46. P. J. Roache, Scaling of High Reynolds Number Weakly Separated Channel Flows, *Proc. Symposium on Numerical and Physical Aspects of Aerodynamic Flows*, California State University at Long Beach, CA, 19-21 Jan. 1981, pp. 87-98.

47. P. J. Roache, Semidirect/Marching Methods and Elliptic Grid Generation, *Proc. Symposium on the Numerical Generation of Curvilinear Coordinate Systems and Use in the Numerical Solution of Partial Differential Equations*, April 1982, Nashville, TN, J. F. Thompson, (ed.), North-Holland Publishing Co., Amsterdam.
48. P. J. Roache, S. Steinberg, and W. M. Money, Interactive Electric Field Calculations for Lasers, AIAA Paper 84-1655, *AIAA 17th Fluid Dynamics, Plasma Physics, and Lasers Conference*, 25-27 June 1984, Snowmass, CO.
49. P. J. Roache, Marching Methods for Elliptic Problems: Part 1, *Numerical Heat Transfer*, Vol. 1, No. 1, 1978, pp. 1-25. Part 2, *Numerical Heat Transfer*, Vol. 1, No. 2, 1978, pp. 163-181.
50. G. H. Golub and C. F. Van Loan, *Matrix Computations*, The Johns Hopkins University Press, Baltimore, 1989.
51. O. Axelsson, *Iterative Solution Methods*, Cambridge University Press, New York, 1994.

Chapter 2

HIGHER-ORDER EQUATIONS

2.1 Introduction

In Chapter 1, we considered only second-order elliptic equations discretized with second- and first-order accurate finite differences. In this chapter, following [1], we will consider "higher-order" operators, both in the sense of *higher-order accuracy* solutions to second-order equations such as the Poisson equation, and in the sense of *higher-order elliptic equations* such as the biharmonic equation. The basic marching method is direct, that is, noniterative, but some of the most powerful techniques presented herein utilize it within rapidly converging iterative schemes.

2.2 Higher-Order Accuracy Operators

It is practical to use $0(\Delta^4)$ and higher-order accuracy finite-difference operators directly in the marching equations, but only in the direction normal to the march direction. As an example, we consider the usual fourth-order analog,

$$\frac{\partial^2\Psi}{\partial x^2} \cong \frac{\delta^2\Psi}{\delta x^2} = \frac{-\Psi_{i+2,j}+16\Psi_{i+1,j}-30\Psi_{i,j}+16\Psi_{i-1,j}-\Psi_{i-2,j}}{12\Delta x^2} \tag{2.2.1}$$

By using the usual three-point second-order analog for $\partial^2\Psi/\partial y^2$, the march for the Poisson equation can proceed in the j direction, replacing Eq. (1.2.5) with

$$\Psi_{i,j+1} = \Delta y^2\left[\zeta_{i,j}-\frac{\delta^2\Psi}{\delta x^2}\right] + 2\Psi_{i,j}-\Psi_{i,j-1} \tag{2.2.2}$$

Adjacent to the left and right boundaries, at $i = 2$ and $i = (I\text{-}1)$, it is necessary to revert to a three-point analog for $\delta^2\Psi/\Delta x^2$, or to change to an uncentered $0(\Delta x^4)$ analog, since Ψ_{i-2} and Ψ_{i+2} are not available. For this mixed three-point, five-point equation, the size of the error term $P = \log_{10}(F_{ic})$ is only $\sim 12\%$ higher than that for the $0(\Delta^2)$ equation (see Appendix A). This might be compensated in a particular problem by taking advantage of the fourth-order accuracy in the x direction to increase Δx and therefore β, which will reduce P.

However, when the analogous $0(\Delta y^4)$ accurate stencil is used in the y (marching) direction, the instability of the march is increased and the method is no longer practical. The increased instability of the $0(\Delta^4)$ method is to be expected, since the higher accuracy means that the discrete analog more closely approaches the Hadamard instability (ill-posedness) of the continuum equation. This also explains the increased instability of the "12" formula for the Laplacian [2]. (See also Appendix A.) Although only $0(\Delta^2)$ accurate for the Poisson equation, it is $0(\Delta^4)$ accurate for the Laplace equation, which *is* the error propagation equation.

The "compact" $0(\Delta^4)$ difference equations [3, 4] are adaptable to a marching solution with no deterioration of stability, but only in one dimension. It can be shown that the $0(\Delta y^2)$ one-dimensional equation

$$\Psi_{j+1} - 2\Psi_j + \Psi_{j-1} = \Delta y^2 \zeta_j \tag{2.2.3}$$

becomes $0(\Delta y^4)$ accurate when ζ_j is replaced by $\tilde{\zeta}_j$, where

$$\tilde{\zeta}_j = \frac{1}{12}\left(\zeta_{j+1} + 10\,\zeta_j + \zeta_{j-1}\right) \tag{2.2.4}$$

Obviously, the error propagation equation is not affected since it is homogeneous. However, in two dimensions, the compact differencing leads to the well-known "20" formula of Bickley (for the case $\Delta x = \Delta y$), a nine-point stencil whose implicit (tridiagonal) march is more than twice as unstable as the $0(\Delta^2)$ stencil (see Fig. 7 of [2]).

The "20" and "12" stencils, as well as the "rotated" or "diagonal unit square" operators, and alternate marching schemes were considered in [2], but none have good enough stability properties to warrant further attention here.

2.3 Higher-Order Accurate Solutions by Deferred Corrections

A simple way to avoid the stability problem of the high-order stencils and still achieve $0(\Delta^4)$ accurate solutions is to iteratively correct the $0(\Delta^2)$ solution to an $0(\Delta^4)$ or higher solution using "deferred corrections" [5]. A direct solution to $\nabla^2\Psi = \zeta$ is obtained with an $0(\Delta^2)$ analog of the ∇^2 operator, giving Ψ^1. From Ψ^1, we calculate at all points a correction term c^1 that is the difference between the $0(\Delta^4)$ and the $0(\Delta^2)$ analogs of $\nabla^2\Psi^1$. (This is somewhat different from the usual description given in [5].) Then a second direct solution of $\nabla^2\Psi = (\zeta + c^1)$ gives Ψ^2, etc. If the $0(\Delta x^2)$ solution is reasonable, these iterations may be expected to converge rapidly, since we are only iterating on the difference between an $0(\Delta^2)$ solution and an $0(\Delta^4)$ solution. Generally, only one or at most two corrections are needed, even when the resolution is poor (four node points per Fourier component), but continued iterations can eventually begin to diverge without relaxation. If one correction is used with the source term being over-written, no storage penalty incurs; otherwise, an additional storage location for $(\zeta + c)$ is required. The one-dimensional five-point correction is not applicable to points adjacent to the boundary points, but significant improvement in accuracy is obtained with the $0(\Delta^4)$ corrections being applied only to interior points farther from the boundary. If the near-boundary points are to be corrected, it is not sufficient [and, in fact, the $0(\Delta^2)$ solution is degraded] to use uncentered five-point equations, which for second-derivative analogs are only $0(\Delta^3)$; it is necessary to correct with uncentered six-point equations that are also $0(\Delta^4)$. The correction terms from the Poisson equation with Dirichlet boundary conditions are

$$c_i^1 = \frac{1}{12\Delta x^2}\left(\Psi_{i-2} - 4\Psi_{i-1} + 6\Psi_i - 4\Psi_{i+1} + \Psi_{i+2}\right) \ \textit{for}\ 2 < i < I-1 \tag{2.3.1a}$$

$$c_2^1 = -\frac{1}{12\Delta x^2}\left(\Psi_1 - 6\Psi_2 + 14\Psi_3 - 16\Psi_4 + 9\Psi_5 - 2\Psi_6\right) \ \textit{for}\ i = 2. \tag{2.3.1b}$$

Analogous terms apply for $i = (I\text{-}1)$, and for the y direction. Near corners, for example, at (2, 2), both the x and y corrections apply. Fourth-order gradient boundary conditions should be set in the direct solution itself, rather than being iterated.

Within semidirect solutions of nonlinear equations (Chapter 8), the $0(\Delta^4)$ corrections need not be made until the $0(\Delta^2)$ solution has approached iterative convergence. Thus, an $0(\Delta^4)$ solution will possibly incur only a 10-20% time penalty compared with the $0(\Delta^2)$ solution. However, we have experienced difficulty in achieving convergence of the deferred corrections for $0(\Delta x^6)$ solutions.

It is also possible to achieve fourth-order accuracy on a subgrid using Richardson extrapolation (e.g., see [6,7]). In [8], we demonstrated the use of the marching methods in a systematically refined grid (11×11 to 81×81) with a 9-point operator, and demonstrated that the very small residual errors of the marching methods allow Richardson extrapolation to be used on the *nonorthogonal* mesh to achieve $0(\Delta^4)$ accurate solutions.

It is worth commenting here on an often misunderstood aspect of numerical methods. High order methods do not solve the cell Reynolds number oscillations, and are no substitute for resolution. If the resolution is adequate to achieve reasonable accuracy, high-order methods can be used to achieve high accuracy solutions more efficiently than just increasing the resolution with second-order methods. But adequate resolution is still required, and in fact often is more important for high order methods.

2.4 Higher-Order Elliptic Equations

As a prototype of higher-order elliptic equations, we consider the biharmonic equation

$$\nabla^4 \Psi = q \tag{2.4.1}$$

where $q = 0$ and the slightly more general nonhomogeneous problem with $q \neq 0$. The $0(\Delta^2)$ analog for ∇^4 is a thirteen-point stencil that extends to $\Psi_{i\pm2,\, j\pm2}$ (e.g., see [9,10] or Chapter 8). We consider three different approaches. First, following Dietrich [11], it is possible to solve the discretized single Eq. (2.4.1) by direct marching. Alternately, we can write Eq. (2.4.1) as two coupled Poisson equations

$$\nabla^2 \Psi = \zeta \tag{2.4.2a}$$

$$\nabla^2 \zeta = q \tag{2.4.2b}$$

with coupled boundary conditions, as in the fluid dynamics problem where Ψ is the stream function and ζ is the vorticity (e.g., see [6]). By using the two-equation formulation, the boundary coupling may be approached directly as by Dietrich [11,12] or iteratively as by Ehrlich [13].

In the two-equation approach as given (for a different system of equations) by Dietrich [12], the boundary coupling of Ψ and ζ is done directly. The lower boundary is set at $j = 1$, where $\Psi_{i,1}$ is given. Then $\Psi'_{i,2}$ and $\zeta'_{i,2}$ are guessed; these are initially in error by $e_{i,2}$ and $e\zeta_{i,2}$, respectively. Then the Poisson equation Eq. (2.4.2a) for Ψ can be marched to $\Psi'_{i,3}$. With $\Psi'_{i,2}$, $\Psi'_{i,3}$, $\Psi_{i,1}$, and $\zeta'_{i,2}$ all known provisionally, the value of $\zeta'_{i,1}$ on the lower boundary may be calculated to $0(\Delta^2)$ accuracy from the given boundary condition on $\partial\Psi/\partial y\,|_{i,1}$ using Woods' or Jensen's equations [6]. The initial error vector $\boldsymbol{E}$ is made up of the elements $e_{i,2}$ and $e\zeta_{i,2}$, so

$\boldsymbol{E}$ is $2L$ long, twice as long as $\boldsymbol{E}$ for the single Poisson equation. The elements of $\boldsymbol{E}$ should be ordered as

$$\begin{aligned} E_m &= e_{i+1,2} \quad \textit{for } m = 1,3,5, \ldots \textit{ with } i = \frac{m+1}{2} \\ E_m &= e\zeta_{i+1,2} \quad \textit{for } m = 2,4,6, \ldots \textit{ with } i = \frac{m}{2}. \end{aligned} \tag{2.4.3}$$

This ordering concentrates the largest terms of C near the diagonal, minimizing round-off error accumulation and aiding the banded approximations given in Chapter 4. Corresponding boundary treatments are used at $j = J$ to define the final error vector $\boldsymbol{F}$.

The procedure for the single biharmonic equation is similar, with two lines of initial values being set by known boundary conditions (as in [9]) and two lines being guessed, so that $\boldsymbol{E}$ is again $2L$ long.

In the two-equation approach using iterative boundary coupling of Ψ and ζ, we follow the LAD method of Chapter 8 (with Re $= 0$). Initial values for the boundary values ζ_b^1 are guessed. Equation (2.4.2b) is solved for ζ^1 at interior points, and Eq. (2.4.2a) is solved for Ψ^1 at interior points. Then the new iteration for the boundary values ζ_b^2 can be solved by a function $f(\zeta^1, \Psi^1)$, where f is some $0(\Delta^2)$ equation such as that of Woods, Jensen, or Israeli [3,6,10]. We have used only the simplest direct substitution type of iteration for ζ_b. However, the iterations do not converge unless the ζ_b's are under-relaxed by some factor $\rho < 1$. (For alternative schemes for the biharmonic equation, see Erlich [13] and references therein and Bjorstad [14].)

2.5 Operation Counts for Higher-Order Systems

The detailed derivation of the operation counts for these different approaches will be presented, closely following the analysis in Chapter 1 for the single Poisson equation. The operation count is significantly increased, both from the unavoidable increased complexity of the equations at each point, and from the doubled length of $\boldsymbol{E}$ and $\boldsymbol{F}$ for the direct coupling methods.

For the 2-equation approach with iterative boundary coupling, each pair of solutions just requires twice the operations of a single Poisson solution. For a problem with $L = M$, from Eq. (1.2.34a) we obtain $\theta_{init} = 8M^3$ plus lower order terms. From Eq. (1.2.34b), we obtain $\theta_{rep} = 28M^2$ for each iteration, or $\theta_{rep} \simeq 28(1+\rho)M^2$ for ρ iterative corrections.

For the 2-equation method with *direct* boundary coupling, the operation count is more involved. Each march for the pair Eqs. (2.4.2) involves one homogeneous equation with 2* and 3$\pm$ at each point, and one non-homogeneous equation 2* and 4$\pm$ at each point. The generation of C now requires $2L$ marches (one for each $\boldsymbol{e}_i = 1$ and $\boldsymbol{e}\zeta_i = 1$ for $2 \le i \le L$) over LM points. These are reduced by a factor λ by null calculations, approximately the same as the one-equation factor. This gives a total θ_2 to establish C,

$$\begin{aligned} \theta_2 &= 8\lambda L^2 M *, \ 14\lambda L^2 M \pm \\ &= 22\lambda L^2 M \ \textit{operations} \end{aligned} \tag{2.5.1}$$

which is slightly over 4 times the operations required in Eq. (1.2.26a) to establish C for one Poisson equation solution. The LU decomposition of C requires θ given by Eq. (1.2.27) with $n = 2L$.

$$\theta_3 = \left[\frac{8}{3}L^3 + 2L^2\right] * \textit{ and } \pm, \ 2L^2 \div \tag{2.5.2}$$

This contrasts to $n = L$ for the one-equation solution, increasing this contribution to θ by a factor of 8. The total θ for initialization of the two-equation direct coupling approach is then $\theta_2+\theta_3$ or

$$\theta_{init,DC} = \left[8\lambda L^2 M + \frac{8}{3}L^3 + 2L^2\right] * \quad \left[14\lambda L^2 M + \frac{8}{3}L^3 + 2L^2\right] \pm, \quad \left[2L^2\right] \div \tag{2.5.3}$$
$$= 22\lambda L^2 M + \frac{16}{3}L^3 + 6L^2 \quad operations$$

Repeat solutions require two marches of the pair of equations, with 2*, 4$\pm$ and 2*, 3$\pm$ operations respectively, plus the repeat (backsolve) LU solution from Eq. (1.2.32), again with $n=2L$. The result is

$$\theta_{rep} = \left(4L^2 + 8LM\right)*, \quad \left(4L^2 + 14LM\right)\pm \tag{2.5.4}$$
$$= 8L^2 + 22LM \quad operations$$

This is increased by only $2L\pm$ for the non-homogeneous biharmonic equations.

Finally, we consider the approach using the single biharmonic equation (2.4.1), discretized with $0(\Delta^2)$ accuracy [9,10]. For the special case of $\Delta x = \Delta y$ and $q = 0$, we obtain the formula

$$D^4\Psi_{i,j} = 0 \tag{2.5.5}$$

The more general case can be written as

$$\begin{aligned} D^4\Psi_{i,j} &= \Psi_{i-2,j} + \Psi_{i,j+2} + \Psi_{i+2,j} + \Psi_{i,j-2} \\ &+ 2\left[\Psi_{i-1,j+1} + \Psi_{i+1,j+1} + \Psi_{i+1,j-1} + \Psi_{i-1,j-1}\right] \\ &- 8\left[\Psi_{i-1,j} + \Psi_{i,j+1} + \Psi_{i+1,j} + \Psi_{i,j-1}\right] \\ &+ 20\Psi_{i,j} \end{aligned} \tag{2.5.6}$$

$$D^4\Psi_{i,j} = \frac{D^2\Psi_{i+1,j} - 2D^2\Psi_{i,j} + D^2\Psi_{i-1,j}}{\Delta x^2} + \frac{D^2\Psi_{i,j+1} - 2D^2\Psi_{i,j} + D^2\Psi_{i,j-1}}{\Delta y^2} \tag{2.5.7a}$$

where

$$D^2\Psi_{\ell,m} = \frac{\Psi_{\ell+1,m} - 2\Psi_{\ell,m} + \Psi_{\ell-1,m}}{\Delta x^2} + \frac{\Psi_{\ell,m+1} - 2\Psi_{\ell,m} + \Psi_{\ell,m-1}}{\Delta y^2} \tag{2.5.7b}$$

When Eq. (2.5.7) is expanded, and cell factors like $(\Delta x/\Delta y)^2$ etc., are pre-stored, we count 8*, 18± for each application of the $0(\Delta^2)$ approximation to the biharmonic operator. We again require $2L$ marches over L x M points reduced by a factor λ by null calculations. This gives

$$\begin{aligned}\theta_2 &= 16\lambda L^2 M *,\quad 36\lambda L^2 M \pm \\ &= 52\lambda L^2 M \quad \textit{operations}\end{aligned} \tag{2.5.8}$$

to establish C. This is about 2.4 times the operations to establish C for the two-equation direct coupling approach, and about 10.4 times that for a single Poisson equation from Eq. (1.2.26a). The LU decomposition is the same as that for the two-equation direct coupling, given by Eqs. (2.5.2). The total θ for initialization of the single biharmonic equation is then $\theta_2+\theta_3$ or

$$\begin{aligned}\theta_{init} &= \left[16\lambda L^2 M + \frac{8}{3}L^3 + 2L^2\right] *, \\ &\qquad \left[36\lambda L^2 M + \frac{8}{3}L^3 + 2L^2\right] \pm,\ \left[2L^2\right]\div \\ &= 52\lambda L^2 M + \frac{16}{3}L^3 + 6L^2 \quad \textit{operations}\end{aligned} \tag{2.5.9}$$

Repeat solutions require two marches, each with 8*, 18±, plus the repeat LU solution as in Eq. (1.2.31) with $n = 2L$. The result is

$$\begin{aligned}\theta_{rep} &= \left(4L^2 + 16LM\right) *,\ \left(4L^2 + 36LM\right) \pm \\ &= 8L^2 + 52LM \quad \textit{operations}\end{aligned} \tag{2.5.10}$$

The marching method for the biharmonic equation thus has an operation count for repeat solutions that is "optimal," i.e., merely proportional to the number of unknowns.

Note that we have *not* made use of the possible symmetry of C in these operation counts (see Section 1.2).

The results are summarized in Table 2.5.1 for the case $L = M$, for which $\lambda = 2/3$. The single-equation biharmonic solution is seen to have *no* advantage over the two-equation approach using direct coupling at the boundaries; besides being about a factor of 2 slower in both initialization and repeat solutions, it gives a value of F_{lc} about 10 times as large for $\Delta x > \Delta y$, meaning that the algebraic solution loses about one more decimal significant figure of accuracy. Also, the truncation error is generally larger, although still $0(\Delta^2)$; see [9].

Point of Comparison	Single equation, $\nabla^4\Psi = q$	Direct Coupling, $\nabla^2\Psi = \zeta,\ \nabla^2\zeta = q$	Iterative Coupling, $\nabla^2\Psi = \zeta,\ \nabla^2\zeta = q$
Storage required for C's, for L×M internal points	$4L^2$	$4L^2$	$2L^2$, or L^2 for simplest case
Truncation error	worse, still $0(\Delta^2)$	basis	basis
Algebraic accuracy	worse than iterative by ≃ 1 figure	somewhat worse than iterative	basis
θ_{init}, L = M	$40\ M^3$	$20\ M^3$	$8\ M^3$
θ_{rep}, L = M	$60\ M^2$	$32\ M^2$	$28(1+r)\ M^2$

Table 2.5.1. Comparison of three approaches for solving the nonhomogeneous biharmonic equation by marching methods. Only the leading terms are given for the operation counts θ, and symmetry of the influence coefficient matrix has not been utilized. Each *, ÷, or ± is counted as an operation. "Basis" indicates that entry is the standard of comparison. From Roache [1].

The remaining choice is between direct coupling or iterative coupling of the two equations. Since the iterative coupling approach has a C that is only half as long as in the direct coupling approach, it requires less storage and will be more accurate algebraically in the Gaussian elimination. The choice will depend on several factors, with the iterative coupling approach favored by increasing problem size, decreasing number of repeat solutions, nesting within a semidirect solution (Chapter 8), and of course, by the lower number of required iterative corrections r and the availability of an estimate for optimum ρ. In our experience, $(1+r)$ has ranged from 6 to 30 for optimum values of ρ. For example, if $(1+r) = 6$ and $M = 100$, we find the iterative coupling is faster for ~8 or less repeat solutions, but for $(1+r) = 30$ and $M = 50$, we find that iterative coupling is not as fast as direct coupling even for one repeat solution. Iterative coupling is favored strongly as the order of the elliptic system increases; for a very high order system, storage considerations alone will dictate the use of iterative coupling. Also, if iterative coupling is already used to extend the problem size via Domain Decomposition (Chapter 3), then iterative coupling of the Ψ and ζ equations is naturally efficient.

2.6 Finite Element Equations

We have briefly considered application of marching methods to some finite element discretizations on a regular mesh. The difficulty is that many finite element discretizations, say for the Poisson equation, are fourth order accurate for the homogeneous problem. Even though the problem being solved may be nonhomogeneous, the marches to establish the influence coefficient matrix are homogeneous and therefore higher order accurate and highly unstable, as noted in Section 2.2. The only approach via marching methods would have to be through a deferred correction, where a low order, possibly finite difference, discretization is solved directly by marching methods, and the final finite element discretization is solved by iteration. This would seem to be possible only for a regular grid structure.

References for Chapter 2

1. P. J. Roache, Marching Methods for Elliptic Problems: Part 2, *Numerical Heat Transfer*, Vol. 1, 1978, pp. 163-181.
2. P. J. Roache, A Direct Method for the Discretized Poisson Equation, Sandia National Laboratories, Report SC-RR-70-579, Albuquerque, NM, 1971.
3. S. A. Orszag and M. Israeli, Numerical Solution of Viscous Incompressible Flows, *Annual Review of Fluid Mechanics*, Vol. 6, 1974, pp. 281-318.
4. R. S. Hirsh, Higher Order Accurate Difference Solutions of Fluid Dynamics Problems by a Compact Differencing Scheme, *Journal of Computational Physics*, Vol. 19, 1975, pp. 90-109.
5. L. Fox, *The Numerical Solution of Two-Point Boundary Value Problems in Ordinary Differential Equations*, Clarendon Press, Oxford, England, 1957.
6. P. J. Roache, *Computational Fluid Dynamics,* rev. printing, Hermosa Publishers, Albuquerque, NM, 1976.
7. P. J. Roache, A Method for Uniform Reporting of Grid Refinement Studies, ASME FED-Vol. 158, *Quantification of Uncertainty in Computational Fluid Dynamics*, ASME Fluids Engineering Division Summer Meeting, Washington, DC, 20-24 June 1993. I. Celik, C. J. Chen, P. J. Roache, and G. Scheurer,. eds., pp. 109-120. See also P. J. Roache, A Method for Uniform Reporting of Grid Refinement Studies, *Proc. 11th AIAA Computational Fluid Dynamics Conference*, July 6-9, 1993, Orlando, FL, Part 2, pp. 1-57-1058. See also P. J. Roache, Perspective: A Method for Uniform Reporting of Grid Refinement Studies, *ASME Journal of Fluids Engineering*, Vol. 116, Sept. 1994, pp. 405-413.
8. P. J. Roache, Scaling of High Reynolds Number Weakly Separated Channel Flows, *Proc. Symposium on Numerical and Physical Aspects of Aerodynamic Flows*, California State University, Long Beach, CA, Jan. 19-21, 1981.
9. P. J. Roache and M. A. Ellis, The BID Method for the Steady-State Navier-Stokes Equations, *Computers and Fluids*, Vol. 3, 1975, pp. 305-320.
10. P. J. Roache, The Split NOS and BID Methods for the Steady-State Navier-Stokes Equations, in R. D. Richtmyer (ed.), *Lecture Notes in Physics, Vol. 35, Proc. Fourth International Conference on Numerical Methods in Fluid Dynamics*, Springer, New York, 1975, pp. 347-352.
11. D. E. Dietrich, private communication, 1977.
12. D. E. Dietrich, Numerical Solution of Fully Implicit Energy Conserving Primitive Equations, *Journal of the Meteorological Society of Japan*, Vol. 53, No. 3, 1975, pp. 222-225.
13. L. W. Ehrlich, Point and Block SOR Applied to a Coupled Set of Difference Equations, *Computing*, Vol. 12, 1974, pp. 181-194.
14. P. Bjorstad, Fast Numerical Solution of the Biharmonic Dirichlet Problem on Rectangles, *SIAM Journal on Numerical Analysis*, Vol. 20, No. 1, 1983, pp. 59-71.

Chapter 3

EXTENDING THE MESH SIZE: DOMAIN DECOMPOSITION

3.1 Introduction

The major limitation to the usefulness of marching methods for elliptic equations is their inherent instability, which limits the size of the problem that can be solved on a given computer. As described in Chapter 1, this limitation is favorably affected by a large cell aspect ratio $\beta = \Delta x/\Delta y$; for $\beta > 1$, fairly large problems can be solved. However, for $\beta = 1$, the problem size in the j (marching) direction on a computer with 14 significant figures of floating point accuracy is limited to about $J = 16$, giving 4 or 5 significant figures of accuracy.

In this chapter, we consider several direct and iterative methods for extending the mesh size, that is, for increasing the maximum J of the problem for a given computer word length, thereby overcoming the inherent instability of the marching method.

Several of these methods are now described as Domain Decomposition methods, although they were developed and published [1] before the term "domain decomposition" came into use. These include both direct coupling of the domains (in domain decomposition terminology, by the "Schur complement matrix") or iterative coupling, either with non-overlapping subdomains (often described as "substructuring") or with overlapping subdomains (as in the "Schwarz alternating procedure"). As such, these methods as described are applicable to other than marching methods. In fact, the original Schwarz alternating procedure dates to the 19th century [2,3] and was used with analytical solutions to Laplace equations on subdomains to prove existence on composite domains. The *numerical* Schwarz alternating procedure was first applied [4] by K. Miller [5] in 1965, but its potential was not widely recognized until the 1980's, e.g., see Glowinski et al. [6] and Chan et al. [7]. It is especially suitable for multi-processor parallel computation; see Brochard [8], Haghoo and Proskurowski [9], and especially Gropp and Keyes [10]. Other domain decomposition methods described herein and in Chapter 4 are more intimately tied in to the marching methods.

Note, however, the differing motivations. The current popularity of Domain Decomposition methods (there are many) is motivated primarily either by the natural domain structuring of problems or the need to parallize algorithms for large problems, whereas originally [1] it was stabilization.

The natural domain structuring can include geometric composition (e.g., T-shaped regions decomposed into rectangular blocks, either overlapping or non-overlapping, or boundary-fitted grid modules for an airplane fuselage, wings, vertical tail, horizontal tail, external stores, etc., again either overlapping or non-overlapping), and governing physical equations (e.g., free-stream potential flow, Euler equations, laminar viscous boundary layers, transitional and turbulent viscous flow), and discretized equations (e.g., unstructured finite element discretization patched to structured finite difference or finite volume equations, low-order and high-order discretizations, non-orthogonal grids for external flows patched to orthogonal grids for boundary layers). Although these "natural domain structuring" problems are intuitively and aesthetically appealing, their payoff in terms of increased efficiency is not as large as one might expect at first glance, because the computational load tends to be concentrated in one or two of the domains. For example, turbulence calculations with a multi-equation turbulence model have both more complex equations (more degrees of freedom) and higher resolution requirements than free-stream flows. In the (wide) experience of Lohner and Morgan [11], the gains usually are limited to factors of 2-4 on serial computers. Although not insignificant, these are disappointing in terms of long term needs, and will be reduced further on multiprocessor parallel computers because load balancing (amongst the processors) will be impractical.

The more practical motivation for domain decomposition is the need to parallize algorithms for large problems. The definition of "large" is of course time-dependent, since "large" problems of the 1970's are now solved quickly on workstations or even personal computers, due to both increases in hardware speed and algorithm efficiency (including the algorithms of this book). The serial speed of computers is inherently limited by miniaturization limits and the speed of light, and it is now widely accepted that the "large" problems by 1990's standards will require large- to massively-parallel machines, say from 64 to 1024 multiprocessors. These architectures require special algorithms in order to achieve high parallel efficiency, i.e., to be making *useful* calculations with almost all of the processors almost all of the time.

In evaluating candidate algorithms, it is essential to recognize that there often is a trade-off between *algorithmic efficiency* and *machine (implementation) efficiency* on parallel architectures. As an elementary example, consider the two oldest point-relaxation methods for the Poisson equation, Jacobi (also known as Richardson) iteration and Gauss-Seidel (also known as Liebman) iteration. The former is a natural parallel algorithm, while the latter is essentially serial. By choosing the algorithm naturally suited for a parallel machine, the analyst gives up a factor of 2 in convergence speed [12]. (Red-black ordering of Gauss-Seidel reclaims most of the algorithmic advantage for large problems, but now only half of the problem can be processed in parallel.) This simple example is in fact applicable to block relaxation in domain decomposition [4,13].

Herein, the primary motivation for the use of domain decomposition is to extend the mesh size, i.e., to stabilize the marching methods by reducing the size of the directly-solved problems. From the viewpoint of algorithmic efficiency, the number of subdomains should be minimized, compatible with the requirement of stability. This optimum "granularity" is dependent on the discretized problem and on the computer word length, and has nothing to do with the optimum granularity for parallel processing. Fortunately (for the marching methods), the granularity of present multiprocessor parallel computers already outweighs the granularity requirements of marching methods for many problems, meaning that no additional stabilization is required. When it is, the domain decomposition for stabilizing the marching method can be applied *within* a subdomain designed to meet multiprocessor constraints. The optimal mix will be problem- and machine-dependent, but the basic concepts described herein will be applicable to both the "inner" and "outer" domain decompositions.

3.2 Mesh Doubling by Two-Directional Marching

Hirota et al. [14] presented a method of coupling J by marching in both directions. In Chapter 1, we have described their GSM marching algorithm, which differs only slightly from the basic EVP algorithm shown schematically in Fig. 1.2.1. In the EVP algorithm, the final error vector F is defined by (true) Ψ - (provisional) Ψ' at $j = J$; in the GSM algorithm, F is replaced by the final residual vector $\mathbf{R}$ evaluated at $j = (J - 1)$ by Eq. (1.2.40).

The maximum J can be doubled by defining the elements of the $\mathbf{E}$ vector e, both at $j = 2$ and at $j = (J - 1)$, and the elements of the $\mathbf{R}$ vector r along two adjacent lines in the center of the mesh (see Fig. 3.2.1b). Suppose $\beta = 1$ and the computer is a Cray, so that $J = 16$ is the largest value that gives satisfactory accuracy for a single march direction. We apply this mesh-doubling method using $J = 32$. We guess provisional values of $\Psi'_{i,2}$ and $\Psi'_{i,31}$ for all i. The discretized Poisson equation (Eq. 1.2.15) is solved for $\Psi'_{i,j+1}$ and marched in the form of Eq. (1.2.17) from $j = 2$ to $j = 15$, giving residuals $= 0$ through $j = 15$ and giving provisional values at $\Psi'_{i,16}$. From the other side of the mesh, the discretized Poisson equation is solved for $\Psi'_{i,j-1}$ and marched, in the form of Eq. (1.2.17) with subscripts $(j \pm 1)$ interchanged, from $j = (J-1) = 31$ to $j = 18$, giving residuals $= 0$ down through $j = 18$ and giving provisional values at $\Psi'_{i,17}$. The nonzero residuals r along $j = 16$ and 17 make up the residual vector R_l. The size of the influence matrix C is now enlarged from LxL to $2Lx2L$. No further characteristics are given in [14], but we present here a detailed operation count [1] for the Poisson equation, following Chapter 1.

Initialization requires the establishment of the matrix C by marching the homogeneous equation, and the LU decomposition of C. If null calculations are *not* avoided, the $2L$ marches over $L\,(M/2)$ points would give the same contribution to the operation count as the single-direction algorithm, given by Eq. (1.2.26a) with $\lambda = 1$. However, *more* null calculations exist in the double-direction algorithm. For $L = M$, Eq. (1.2.26b) gives $\lambda = ⅔$ for the single-direction algorithm, but it may be verified that the two-directional marching algorithm gives $\tilde{\lambda} = ⅓$, reducing this contribution by a factor of ½. (If crossderivatives are present, both give $\lambda = 1$; see the Section 1.3.6 on cross derivatives in Chapter 1.) However, the contribution from the LU decomposition is increased because **E** and **R** are twice as long, so that $n = 2L$ in Eq. (1.2.27).
The result is

$$\theta_{init,dbl} = \left(2\tilde{\lambda}L^2M + \tfrac{8}{3}L^3 + 2L^2\right)*, \left(3\tilde{\lambda}L^2M + \tfrac{8}{3}L^3 + 2L^2\right)\pm, \left(2L^2\right)\div \tag{3.2.1}$$

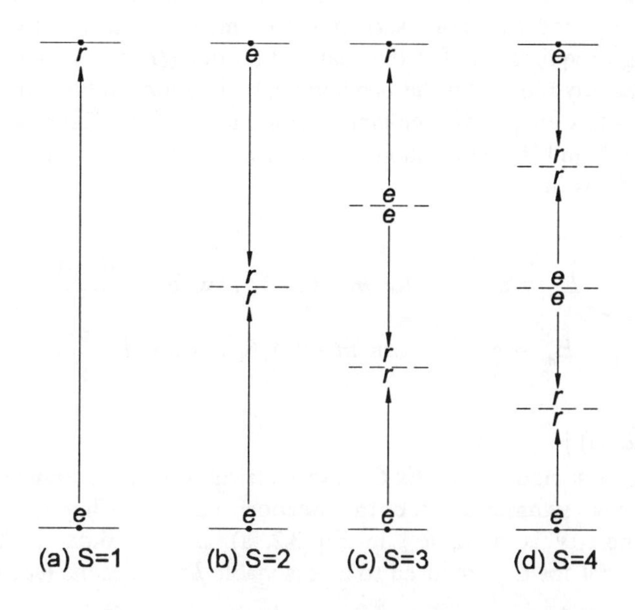

Figure 3.2.1. Multiple Marching methods for S subregions. Shown are the positions of rows of e and r, the elements of the initial and final error vectors E and R. Dashed lines separate the subregions between nodes. The elliptic equation is applied at e and up to, but not including, the r positions, in the direction indicated by the arrows. For example, the elliptic equation might be satisfied at $j = 2, 3, 4, ..., 15, 16$, which gives $\Psi'_{i,17}$ and $r_{i,17} \neq 0$.

Repeat solutions require two marches of the Poisson equation (or the residual evaluation) over all interior points, and the solution for **E**. The first part is unchanged from the single-direction algorithm, and the second requires $n = 2L$ in Eq. (1.2.32). The result is

$$\theta_{\text{rep},dbl} = (4LM + 4L^2) *, (8LM + 4L^2) \pm \tag{3.2.2}$$

Counting "operations," Eqs. (3.2.1) and (3.2.2) become

$$\theta_{\text{init},dbl} = 5\,\tilde{\lambda}L^2M + \tfrac{16}{3}L^3 + 6L^2 \tag{3.2.3a}$$

$$\theta_{\text{rep},dbl} = 12LM + 8L^2 \tag{3.2.3b}$$

Specializing to $L = M$, for which $\tilde{\lambda} = 1/3$ for the five-point equation, we obtain

$$\theta_{\text{init},dbl} = 7M^3 + 6M^2 \tag{3.2.4a}$$

$$\theta_{\text{rep},dbl} = 20M^2 \tag{3.2.4b}$$

[For $\tilde{\lambda} = 1$, Eq. (3.2.4a) would give $\theta = 10\tfrac{1}{3}M^3 + 6M^2$.]

Comparing Eq. (3.2.4) with Eq. (1.2.34), we see that the need for using two-directional marching to extend the mesh size involves a penalty of about 75% in initiation time and 43% in repeat solution time for the Poisson equation. This is a reasonable penalty, and the more serious requirement of the method is the fourfold increased storage requirement for the matrix C, from $((L \times L)$ to $(2L \times 2L)$. This penalty will be greatly reduced if the banding approximations to be covered in Chapter 4 can be used. To this end, and to enhance the accuracy of the Gaussian elimination for the matrix C, the **E** and **R** vectors should be ordered so as to keep the largest terms near the diagonal of C, as in

$$E_m = e_{i+1,2} \quad \text{for } m = 1,3,5, \ldots \text{ with } i = \frac{m+1}{2} \tag{3.2.5a}$$

$$E_m = e_{i+1,J-1} \quad \text{for } m = 2,4,6, \ldots \text{ with } i = \frac{m}{2}. \tag{3.2.5b}$$

[See also Eq. (2.4.3).]

We have not assumed the matrix C is symmetrical, which can easily occur even if the solution Ψ is not symmetric (see the Section 1.2.4 in Chapter 1 on operation count). If symmetry is used, θ_{init} in Eq.(3.2.4a) is approximately halved, and the storage requirement for C is reduced to a reasonable L^2 locations (see Chapter 1).

3.3 Multiple Marching

The method described above can be extended to more than two subregions, schematically represented in Fig. 3.2.1. Part *a* shows the position of the *e* and *r* rows for a single march, and part *b* shows the two rows of *e* and two rows of *r* for the two-directional marching algorithm. (In any of the procedures described here, the indicated positions of *e* and *r* could be interchanged.) Parts *c* and *d* indicate the positions of *e* and

r rows for three and four subregions, respectively; generalizations are obvious to S subregions.

The connection is clear between this multiple marching of linear second-order partial differential equations and the *parallel shooting* methods of Keller [15] for systems of (possibly) nonlinear first-order ordinary differential equations. In Keller's method, the first-order equations require only a single row of e at each subregion boundary, whereas the present method with second-order equations requires pairs of guessed values Ψ' at the e locations in order to initiate the independent multiple marches. (If single rows of e rather than pairs were used, the marches in the subregions would not be independent and the size of Ψ_J' would be reduced only by the equivalent of reducing J by $2S$, which is not adequate.)

The operation count will depend strongly on L, M, S, and λ, but the natural limiting case to consider is S large. The length of the march in j within each subregion is M/S. For large S, $L/(M/S)$ is large and the number of null calculations increases and λ becomes small. (For example, in the limit of maximum S, the rows starting from $i = 2$ would be $e, r, r, e, e, r, r, \ldots$ and we have $\lambda = 1/L$). Thus the contribution to θ_{init} from the error propagation equation becomes insignificant [see Eq. (1.2.26a)] compared with the contribution from the LU decomposition. For S subregions, the **E** vector is SL long (see Fig. 3.2.1). As in Eq. (1.2.27), the LU decomposition of an $n \times n$ matrix requires

$$\theta_{LU} = \left(\frac{1}{3}n^3 + \frac{1}{2}n^2\right) *, \left(\frac{1}{3}n^2 + \frac{1}{2}n^2\right) \pm, \left(\frac{1}{2}n^2\right) \div \tag{3.3.1a}$$

and the back solution requires

$$\theta_{BS} = \left(n^2\right) *, \left(n^2\right) \pm \tag{3.3.1b}$$

When $n = SL$ in Eq. (3.3.1a), and counting operations, we obtain

$$\theta_{init,S \gg 1} \cong \frac{2}{3}S^3L^3 + \frac{3}{2}S^2L^2 \tag{3.3.2a}$$

For repeat solutions, we have two marches that give approximately θ_4 from Eq. (1.2.30). [Actually, in the limit of maximum S, half of these θ_4 operations would be replaced by a simple $\pm$ in the solutions for $\Psi = \Psi' + e$, but this is not a realistic case to consider, so we keep the conservative estimate of θ_4 from Eq. (1.2.30).] The LU repeat solution gives θ_{BS} from Eq. (3.3.1b) with $n = SL$. Counting operations, we get

$$\theta_{\text{rep},S \gg 1} \cong 12LM + 2S^2L^2 \tag{3.3.2b}$$

Comparing Eqs. (3.3.2) for $L = M$ with Eqs. (1.2.34) for a single region, we see that the penalty ratio with large S is about $1/6\ S^3$ for initiation and about $1/7\ S^2$ for repeats. Furthermore, the difficulty in solving the matrix C by Gaussian elimination increases since the number of equations is equal to SL, and the storage problem gets out of hand since C is $SL \times SL$. For these reasons, the multiple marching method is probably limited to $S = 4$ or 5 for practical purposes, unless banding approximations are used (see Chapter 4).

Note that the multiple marching scheme is well suited for parallel processing computers (as pointed out by Dietrich [16,17]) since many marches can be performed simultaneously.

If the matrix C is symmetric, the θ_{init} will be reduced by roughly 1/2, and the storage for C by a factor of 1/4. See Chapter 1.

3.4 Patching

The technique given here is a *capacity matrix* technique or *influence coefficient* method, which is not connected essentially to marching methods. In fact, the simple and lucid description of it for two subregions was given by Poukey et al. in [18], not in connection with marching methods, but as a technique for solving an L- or T- shaped region by patching together solutions on two rectangular regions obtained by Hockney's method (see also Buzbee et al. [19]). In Domain Decomposition terminology, this is a "Schur complement matrix" system; e.g., see [6,7,20].

As in the section on mesh doubling by two-directional marching, we consider first a doubling of J, with two subregions divided by patching interface at $j = p$. We pick arbitrary values of $\hat{\Psi}$ along the interface, perhaps $\hat{\Psi}_{i,p} = \Psi_{1,p}$. Then an elliptic solution is obtained by marching in each subregion. An initial error vector $\hat{\mathbf{E}}_m$ is defined by

$$\Psi_{i,p} = \hat{\Psi}_{i,p} + \hat{\mathbf{E}}_m \qquad m = i-1 \tag{3.4.1}$$

and a final error of residuals $\hat{\mathbf{R}}_l$ is defined as

$$R_l = r_{i,p} = \frac{\hat{\Psi}_{i+1,p} - 2\hat{\Psi}_{i,p} + \hat{\Psi}_{i-1,p}}{\Delta x^2} + \frac{\hat{\Psi}_{i,p+1} - 2\hat{\Psi}_{i,p} + \hat{\Psi}_{i,p-1}}{\Delta y^2} - \zeta_{i,p} \qquad l = i-1 \tag{3.4.2}$$

The linear relation between $\hat{\mathbf{E}}_m$ and $\hat{\mathbf{R}}_l$

$$\hat{\mathbf{R}}_l = \hat{C}_{lm}\hat{\mathbf{E}}_m \tag{3.4.3}$$

is established by solving the homogeneous equation with $e_{m1,jp} = 1$, all other $e_{m,jp} = 0$ to establish the m_1-th column of $\hat{C}$, etc., analogously to Chapter 1.

The operation count θ for this technique is evaluated for the Poisson equation using the results of Chapter 1, Section 1.2.4. All those results are linear in λM for θ_{init} and linear in M for θ_{rep}. For simplicity, we consider the size of the two subregions equal. As in the section on mesh doubling by two-directional marching, the factor $\hat{\lambda}$ is reduced to $\sim$⅓ by the increase in null calculations, since each march in the generation of C is only $\sim M/2$ long. Thus, the initialization and repeats for both subregions (without combining the solutions) require the same θ's as given in Section 1.2.4. After each subregion is initialized, the initialization for $\hat{C}$ requires L repeat solutions in each subregion, plus an LU decomposition with θ from Eq. (1.2.32). Counting operations with $L = M$, we obtain

$$\theta_{\text{init, 2 patch}} \cong 17M^3 \tag{3.4.4a}$$

This is reduced only slightly, to 15-5/6 M^3, if the two subregions are identical in size, type of boundary conditions, and coefficients of the elliptic equation [1]. The largest contribution to Eq. (3.4.4a) comes from the L repeat solutions in each subregion to establish $\hat{C}$; this contribution is not affected by the two subregions being identical, although this would save storage.

The repeat operations require two repeat solutions in each of the subregions, plus a lesser contribution as in Eq. (1.2.32) of $2L^2$ from the LU solution for $\hat{C}$. For $L = M$, this gives

$$\theta_{\text{rep, 2patch}} \cong 30M^2 \tag{3.4.4b}$$

Comparing Eqs. (3.4.4) with Eqs. (1.2.34) and (3.2.4), we see that for two patched regions, θ_{init} is about 4.3 times that for a single region and 2.4 times that for two-directional marching, and θ_{rep} is about 2.1 times that for a single region and about 1.5 times that for two-directional marching. However, the matrix $\hat{C}$ and the C's for each subregion are only $L \times L$ matrices, whereas the C for the two-directional marching is $2L \times 2L$. This is less demanding on the Gaussian elimination routine and, for identical subregions, requires only half the storage penalty of the two-directional marching. This advantage becomes more pronounced if more patching subregions are used.

It is significant that $\hat{C}$ for the patching has an entirely different character than the C's for the marching, since $\hat{C}$ is not the result of a marching of an elliptic equation. (As mentioned, the subregion solutions could be obtained by some other direct method.) In fact, the elements of $\hat{C}$ are of $O(1)$ and $\hat{C}$ is diagonally dominant, even for $\Delta x = \Delta y$. Hence, the banding approximation to be discussed in Chapter 4 is applicable to $\hat{C}$.

By applying this patching procedure recursively, any number of subregions can be patched together. The next doubling, to four subregions, gives

$$\theta_{\text{init},4\text{ patch}} \cong 46M^3 \tag{3.4.5a}$$

$$\theta_{\text{rep},4\text{ patch}} \cong 62M^2 \tag{3.4.5b}$$

For continued doubling, θ_{rep} increases by about a factor of 2, and θ_{init} somewhat faster. The Gaussian elimination remains practical, since all matrices remain $L \times L$.

Marching methods are faster than other direct methods for all elliptic problems, as noted in Chapter 1, but their principal advantage is in the non-separable problem. Chan and Renasco [21] note that "if the operator ... is non-separable, there usually are no fast solvers available ... ." Marching methods meet this qualification, and are especially well suited to the Domain Decomposition treatment of non-separable problems.

3.5 Influence Extending

The following method was devised by the present author [1] and is related to a method used by Russo [22]. The basic equation needed is an extension of the definition of the influence coefficient matrix C. It was defined in Chapter 1 [Eq. (1.2.21)], and is repeated here as

$$\mathbf{F}_l = \mathbf{C}_{lm} E_m \tag{3.5.1}$$

where $\mathbf{F}$ is the final error vector (i.e., the values of $e_{i,j}$) at $j = J$, and $\mathbf{E}$ is the initial error vector at $j = 2$. The errors $e_{i,1}$ were assumed equal to zero since the boundary conditions are known there. A more explicit notation would be $\mathbf{F}_J = \mathbf{C}_{1,J}\mathbf{E}_2$ where $C_{1,J}$ is the influence coefficient matrix for the region extending from $j = 1$ to J. The extension of this to any subregion from $j = \alpha$ to β, *without* assuming $e_\alpha = 0$, is

$$e_\beta = C_{\alpha,\beta} e_{\alpha+1} + C_{\alpha-1,\beta} e_\alpha \tag{3.5.2}$$

With $\alpha = 1$, $\beta = J$, and $e_1 = 0$, this reduces to Eq. (3.5.1). The matrix $C_{\alpha-1,\beta}$ would be defined just as $C_{\alpha,\beta}$ is, with one more march step.

Consider just two subregions, from 1 to $\alpha = p$, and from $\alpha = p$ to $\beta = J$. Then

$$e_p = C_{1,p} e_2 \tag{3.5.3}$$

$$e_{p+1} = C_{1,p+1} e_2 \tag{3.5.4}$$

Combining these gives

$$e_{p+1} = C_{1,p+1} C_{1,p}^{-1} e_p \tag{3.5.5}$$

Writing Eq. (3.5.2) over the second subregion, from p to J, gives

$$e_J = C_{p,J} e_{p+1} + C_{p-1,J} e_p \tag{3.5.6}$$

Using Eq. (3.5.5) in Eq. (3.5.6) gives

$$e_J = \left(C_{p,J} C_{1,p+1} C_{1,p}^{-1} + C_{p-1,J} \right) e_p \tag{3.5.7}$$

which defines $\tilde{C}_{pJ}$ so that

$$e_j = \tilde{C}_{pJ} e_p \tag{3.5.8}$$

Note that the matrices that define $\tilde{C}_{pJ}$ are "short", that is, they extend only over a subregion (or a subregion + one step in j), and therefore will not incur round-off problems.

The technique is to guess $\hat{\Psi}_p$ and solve the elliptic equation in the lower subregion, then to continue the march through p to J, calculate e_J from known boundary values, and invert Eq. (3.5.8) to get e_p. Thus the final value $\Psi_p = (\hat{\Psi}_p + e_p)$ is solved before e_2 is known. Then the two subregions can be solved independently. Nowhere do we try to solve e_2 directly from e_J. Even after the solution is obtained for both subregions, we could not accurately march from $j = 2$ to $j = J$ (presuming that the subregion boundary at p is actually needed, i.e., that the unstabilized march from 2 to J is inaccurate).

The operation count for this patching depends more strongly on the nature of the elliptic equations than the simple one-region marching method, especially when the technique is extended to many subregions. The most favorable situation occurs for constant coefficients, which means that each subregion has identical $C_{\alpha\beta}$. (The identical $C_{\alpha\beta}$ can also arise with variable coefficients that are functions of y only and periodic in y over each subregion, but aside from this somewhat contrived case, the more realistic assumptions is constant coefficients.)

The operation counts for this most favorable case are obtained as follows, retaining only the highest-order terms. By assumption $C_{p,J} = C_{1,p}$ and $C_{p-1,J} = C_{1,p+1}$. These can be obtained economically in the constant coefficient case marching out to establish $C_{1,p}$ and then continuing the marches one line further to obtain $C_{1,p+1}$. With the length of the subregion $=m$, we obtain from Eq. (1.2.29) neglecting $O(L^2)$ terms,

$$\theta_{\text{init,sub}} = 5\lambda L^2 m + \frac{2}{3} L^3 \tag{3.5.9}$$

The λ for each subregion is smaller than the λ for a complete region with $L = M$, giving $\lambda \cong 1/3$.

The calculation to establish $\tilde{C}_{pJ}$, as seen from Eq. (3.5.7) involves a matrix inversion and three matrix multiplications. Since the LU decomposition for $C_{1,p}$ has already been accomplished in the initialization for the subregion, we calculate $C_{1,p}^{-1}$ with an additional

$(4/3\,L^3)^*$ and $(4/3\,L^3)\pm$. (See [23], p. 159. Actually, one could do it in $(L^3)^*$ and $(L^3)\pm$ by taking special account of some zeros in the matrix, but we base our operation count on the use of a well-tested and general matrix solution library routine.) The two matrix multiplications each involve $(L^3)^*$ and $(L^3)\pm$, and the matrix addition is only $O(L^2)$. Thus the establishment of $\tilde{C}_{pJ}$ involves

$$\theta_{pJ} = 6\tfrac{2}{3}L^3 \tag{3.5.10}$$

The LU decomposition of $\tilde{C}_{pJ}$ requires $(⅓L^3)^*$ and $(⅓L^3)\pm$, from Eq. (3.3.1a). The total for the influence extending method is then the sum of Eqs. (3.5.9), (3.5.10), and (3.3.1a), giving

$$\theta_{\text{init,2ext}} = 5\lambda L^2 m + 8L^3 \tag{3.5.11}$$

For $L = M$, we have $m = M/2$ and $\lambda \cong ⅓$, or

$$\theta_{\text{init,2ext}} \cong 8\tfrac{5}{6}M^3 \tag{3.5.12}$$

Repeat solutions require (1) a repeat solution on both subregions with guessed $\hat{\Psi}_p$; (2) the solution for e_p; and (3) a second repeat solution with the correction Ψ_p over both subregions. Using θ's from Eqs. (1.2.30) and (1.2.32), we obtain

$$\theta_{\text{rep,2ext}} = \left(8LM + 5L^2\right)*,\left(16LM + 5L^2\right)\pm \tag{3.5.13}$$

For $L = M$,

$$\theta_{\text{rep,2ext}} = 34M^2 \tag{3.5.14}$$

The θ_{init} from Eq. (3.5.12) is roughly half that for the two-patch method from Eq. (3.4.4a), whereas the θ_{rep} from Eqs. (3.5.14) and (3.4.4b) are comparable. The two-extension method requires one more storage of C than the two-patch method, but "scratch arrays" (slow disk storage) can be used in this case with little impact.

For variable coefficients, the contribution from Eq. (3.5.9) to θ_{init} is doubled so that Eq. (3.5.12) is replaced by

$$\theta_{\text{init,2ext}} \cong 10\tfrac{1}{3}M^3 \tag{3.5.15}$$

which is still only 61% of θ_{init} from the 2-patch method of Eq. (3.4.4a). A total of five separate C's have to be stored, compared with three for the patching method (but these can still be stored in scratch arrays).

The influence extending method generalizes to more subregions with straightforward coding. The values of Ψ along the subregion boundaries are solved successively from the top down. The primary contribution to θ_{init} for each additional subregion is the establishment and inversion of the next $\tilde{C}_{\alpha\beta}$. Even with the variable coefficients, λ decreases as S increases, so the first term in Eq. (3.5.9) becomes negligible. For each additional subregion, we obtain

$$\text{Increment in } \theta_{\text{init,ext}} \cong 6L^3 \tag{3.5.16}$$

The choice among these three direct methods (multiple marching, patching, and influence extending) thus depends on many factors: tolerance of the Gaussianelimination to large matrices, initiation speed, repeat speed, storage penalty, whether the coefficients are variable or constant, the number of required subregions, and coding simplicity. All methods achieve the extension in problem size at the expense of more matrix operations, either larger matrices or more matrices. The methods could also be combined, as in Dietrich [16].

3.6 Other Direct Methods for Extending the Mesh Size

There are several other prescriptions available to extend the usable value of J. As is the case when trying to distinguish various marching methods (see Chapter 1), it is confusing and of questionable profit to try to distinguish one method from another, or a method from an algorithm. These methods all have in common a significant penalty in the operation count, and are often explicitly related to the idea of orthogonalization. Lucey and Hansen [24] are apparently responsible for one of the earliest of these methods for stabilization/orthogonalization (see also [25] for one-dimensional problems), with later versions by Coleman [26] and Bank and Rose [27,28]. Since all the methods produce direct algebraic solutions of linear equations, it is assured that linear algebraic relations exist among the methods themselves, confusing the question of just what *is* a distinct method. However, that the algorithms are truly distinct may be inferred from the limitations that the authors place on them (e.g., Coleman's [26] algorithm cannot handle analogs of first derivatives in the marching direction; Bank and Rose [27-29] require at least separable equations, and do much better with constant coefficients, etc.) and from the differing operation counts. Particularly, the Bank and Rose methods [27-29] are multiple marching methods like the one described in the section on multiple marching, but their methods depend inherently on the use of modified Chebyshev polynomials; this restricts the methods to variables-separable elliptic equations, but also results in an optimal $O(M^2)$ operation count even for *initialization*.

Scott and Watts [30] use a modified Gram-Schmidt orthonormalization procedure in connection with their superposition marching method. This procedure gives an operation count penalty that is just linearly additive with each additional subregion. Other methods for accomplishing the extension that are similar but not identical to the patching described above are given by Dietrich [16] (who actually uses superposition), Madala [31], and Russo [22]. Dietrich [16] also treats the interesting problem of a changing cell aspect ratio [$\beta \equiv \Delta x/\Delta y = f(x, y)$] giving $\beta < 1$ in some regions with marching in y, and $\beta > 1$ in other regions with marching in x. Yee [32] used a method similar to the two-directional march described in the section on mesh doubling by two-directional marching. Berkerman et al. [33] used a stabilized marching method implemented by Baer and Gross [34] to solve the Poisson equation for double-diffusive free convection flows in 100×100 uniform grids and larger. See also Tang's [39] method of wavefront elimination and renormalization. Dietrich has successfully used a stabilized marching method for the direct solution of the hydrostatic pressure equation in the Sandia Ocean Modeling System [35-38], and for diverse applications in high energy gas laser reactors [63], coal gasifier reactors [64], and simulation of nozzle sprays and atomization [65-67].

Any of these direct methods for extending the mesh size could be coupled with the iterative methods, to be discussed immediately below.

3.7 Lower Accuracy Stencils Plus Iteration

The previous considerations (see Section 2.2 on higher-order accuracy operators) of the causes of instability of the higher-order accuracy stencils suggest [1] the construction of a finite-difference equation for $\partial^2\Psi/\partial x^2$ that is (1) *less* accurate than the usual three-point centered difference equation, and (2) has zero weighting of the point (i,j). Such a stencil is readily derived as

$$\frac{\partial^2\Psi}{\partial x^2}\bigg|_i \cong \frac{\partial\Psi/\partial x|_{i+1\frac{1}{2}} - \partial\Psi/\partial x|_{i-1\frac{1}{2}}}{3\Delta x} \cong \left[\frac{\Psi_{i+2} - \Psi_{i+1}}{\Delta x} - \frac{\Psi_{i-1} - \Psi_{i-2}}{\Delta x}\right]\frac{1}{3\Delta x} = \frac{\Psi_{i+2} - \Psi_{i+1} - \Psi_{i-1} + \Psi_{i-2}}{3\Delta x^2} \tag{3.7.1}$$

This stencil is still $O(\Delta^2)$ accurate, but the leading error term is $5/12\ \Delta x^2(\partial^4\Psi/\partial x^4)$, which is five times the leading error for the usual three-point stencil. As anticipated, it greatly improves the error propagation in the marching solution. The maximum J attainable for the Poisson equation, as given in Eq. (1.2.38), is increased by a factor of about 2.2. Furthermore, the analog Eq. (3.7.1) can be corrected to the usual $O(\Delta x^2)$ analog or directly to an $O(\Delta^4)$ stencil by the deferred corrections, as in Section 2.3 on higher-order accurate solutions by deferred corrections.

A similar approach [1] using more familiar equations is to use the usual formula for $\partial_2\Psi/\partial x^2$ applied over $2\Delta x$, effectively doubling the cell aspect ratio for the march, and to use deferred corrections to achieve the solution over Δx. The march equation analogous to Eq. (1.2.17) is then

$$\Psi_{i,j+1} = \Delta y^2\zeta_{i,j} + \left(2 + \tfrac{1}{2}\alpha\right)\Psi_{i,j} - \tfrac{1}{2}\alpha\left(\Psi_{i+2,j} + \Psi_{i-2,j}\right) - \Psi_{i,j-1} + c^1_{i,j} \tag{3.7.2a}$$

where

$$c^1_{i,j} = \frac{1}{4\Delta x^2}\,\delta^4_x\,\Psi^1_{i,j} \tag{3.7.2b}$$

$$\delta^4_x\,\Psi_{i,j} = \Psi_{i+2,j} - 4\Psi_{i+1,j} + 6\Psi_{i,j} - 4\Psi_{i-1,j} + \Psi_{i-2,j} \tag{3.7.2c}$$

and where superscript 1 refers to the deferred correction term evaluated at the previous iteration level. [Note that $\delta_x^4\Psi_{i,j}/\Delta x^4$ is the usual $O(\Delta x^2)$ approximation to $\partial^4\Psi/\partial x^4$.]

Equations (3.7.2) give an iteration to correct the marched solution of accuracy $O[(2\Delta x)^2, \Delta y^2]$ to accuracy $O(\Delta x^2, \Delta y^2)$. For a little extra work for the Poisson equation, we can correct directly to an $O(\Delta x^4, \Delta y^4)$ solution with the iterative correction $c^1_{i,j}$ in Eq. (3.7.2b) replaced by [1]

$$c^1_{i,j} = \frac{1}{3\Delta x^2}\,\delta^4_x\,\Psi^1_{i,j} + \frac{1}{12\Delta y^2}\,\delta^4_y\,\Psi^1_{i,j} \tag{3.7.3}$$

Special correction terms are required near boundaries, as in Eq. (2.3.1b). The iteration for the deferred corrections may be included within an overall iteration such as semidirect iterations for nonlinear equations (Chapter 8). Obviously, the march equation could be further spread out than Eqs. (3.7.2), limited only by iterative convergence.

As described above, Eq. (3.7.2a) is applied at all columns, i. Instead [1], we may apply it only at odd i, giving a first solution on a *submesh* with doubled β for the first step in the iteration. These values of $\Psi_{\text{odd}\ i,j}$ can then be used as boundary values for the even-numbered columns, which are one-dimensional problems solved independently of each other. Although one-dimensional (i.e., a vector of Ψ values), the stencil is still unstable to a simple marching solution. However, it is readily stabilized by one of the techniques described earlier, and the stabilizing "matrix" (e.g., $\hat{C}$ for the patching method in Section 3.3 on multiple marching) is only a scalar. In the next iteration, deferred corrections are used on the odd-numbered columns, etc. The idea is readily extended to more coarse submeshes; for example, we may solve directly on the submesh $i = 3, 6, 9, 12, \ldots$ followed by two-column solutions for $i = 4$ and 5, $i = 7$ and 8, etc., each stabilized by a $\hat{C}$ that is only 2×2. Since this approach reduces the size of C, it would be preferred for large values of I and for 3-D problems (see Chapter 5).

3.8 Iterative Coupling for Subregions

The problem size for marching methods can also be extended by iterative coupling of the subregions. Although not necessarily required for understanding or implementing these iterative methods, a brief discussion of "preconditioning" is relevant, especially to the literature of Domain Decomposition.

3.8.1 Preconditioning

"Preconditioning" of a matrix problem $Ax = b$ involves defining another matrix that is easier to invert and is an approximation [7] in the sense of being "spectrally close" [40] to the original. Often, as in PCG methods (preconditioned conjugate gradients), the objective is to produce an iterative method in which the matrix is symmetric positive definite. (The term "pre-conditioner" comes from a certain formalism in which a matrix pre-multiplier is used. Although a convergence analysis may use the matrix pre-multiplier directly, one does not ordinarily construct the new matrix by determining the correct pre-multiplier, but simply by a subtraction operation, bringing troublesome terms of A such as variable coefficient or cross-derivative terms over to the right-hand side, and lagging them in an iteration, as in various "splitting" methods.)

In these examples, the pre-conditioner is non-geometric. A more unusual preconditioner is the low-order stencil given in Section 3.7, which is a non-geometric preconditioner created only to partially stabilize marching methods. The term Domain Decomposition was originally used in an obvious geometrical sense, as was "block iteration", but these terms are now sometimes generalized to non-geometric splittings of various kinds. (As noted elsewhere, this extension becomes uselessly fuzzy when line relaxation and even point relaxation could be forced into a formal definition of Domain Decomposition.) A survey of (primarily geometric) preconditioners for Domain Decomposition is given by Chan and Renasco [21]; see also Meurant [41].

The subdomain structuring and creation of the Schur complement matrix or capacitance matrix can lead to block direct methods (i.e., the direct coupling methods such as those of Sections 3.4, 3.5, and 3.6, or other fast domain decomposed solvers [20]), or can lead to block iterative methods via (formally) preconditioners. Note that *neither* the Schur complement matrix *nor* its pre-conditioner need be "actualized" in an array nor even in the formulation in order for an iterative method to be implemented. However, it may be instructive to do so.

The preconditioning problem for domain decomposition, *per se*, can be described precisely for the two-subdomain problem following Gropp and Keyes [10]. The original (full) problem is defined by the matrix A, as in $Ax = B$. The matrix is partitioned into subdomains connected by small (lower dimension) interface regions. For a single domain decomposed into two subdomains (1 and 2) connected by an interface (3), the partitioned matrix would look like

$$A = \begin{bmatrix} A_{11} & 0 & A_{13} \\ 0 & A_{22} & A_{23} \\ A_{13}^T & A_{23}^T & A_{33} \end{bmatrix} \tag{3.8.1}$$

where A_{11} and A_{22} come from the interior of the subdomains, A_{33} from *along* the interface, and A_{13} and A_{23} from the interactions between the subdomains and the interfaces. We are interested in various preconditioners for A, based on their efficacy and on their parallel limitations.

The choice of preconditioner is critical in domain decomposition, as with any iterative method. In the context of parallel computing, a major distinction is between preconditioners that are purely local, those that involve neighbor communication, and those that involve global communication.

The design of the subdomains so as to minimize the size of the interface is important for efficient parallel implementation of MIMD computers, especially of the "message passing" type (as contrasted to the shared memory type) so as to minimize the band width requirements for the communications [10]. The objective in true MIMD computers such as the Intel iPSC Hypercube is to minimize the ratio of communication time to computation time by performing intensive computations locally (Fischer et al., [42]). Note, however, that this ratio is somewhat elusive as a measure of algorithmic efficiency, since it is increased by any increase in processor speed.

3.8.2 Block Iterative Relaxation

Instead of setting the residuals along the interface $R_{i,p} = 0$ in Eq. (3.4.2) by a direct matrix solution, one can drive the residuals toward zero iteratively. A block-relaxation process converges faster than point iteration methods because the errors are longer wavelength. Dietrich et al. [17] have exploited this technique successfully in their BIR method.

For a single interface at $j = p$, their BIR technique is described as follows. A line of values $\Psi_{i,p+1}$ just above the interface is guessed. Then the marching method is used to solve the elliptic equation in the lower region, including the line at $j = p$. With these $\Psi_{i,p}$'s as boundary values, the upper region is solved by a marching method. At this step, the only nonzero residuals are $R_{i,p}$ along the interface. Repeated application will give iterative convergence.

Dietrich et al. [17] refine the method further by extrapolating forward in the iteration process for the interface values, using the linear combination of the first two iterates that gives least-squares residuals along the boundary. They also patch together multiple blocks with interfaces along x and y, although in a later version Dietrich [16] found that wide strips in x, traversing the entire mesh $1 \le i \le$ I, improved the speed by about a factor of 4. It is also possible to avoid the solutions within the blocks between the first and last iterations, so that most of the iteration is done only on the boundary residuals [16]. Other improvements on initial values and convergence rates are given in [16].

The iterative BIR technique may be preferable to the direct matrix solution for many subregions, especially if a good estimate of the guessed values is available as in time-dependent applications [16,17], or if the BIR iteration can be included within an overall iteration such as semidirection iterations for nonlinear equations (Chapter 8). The one-dimensional iteration scheme proposed by Dorodnicyn [43] for patching any two solution regions could also be used, but the BIR method seems preferable.

There are, of course, many variations on this basic block-relaxation scheme; see, e.g., [6,7]. The critical distinguishing feature is that the residuals are evaluated on the boundary between the subdomains and are iteratively driven towards zero.

3.8.3 Schwarz Alternating Procedure

In the earliest Domain Decomposition method (19th century!), the iterative Schwarz alternating procedure [2,3], subdomains are overlapped. A straightforward description is given by Ehrlich [4].

Consider two subdomains, A and B, overlapped as shown in Figure 3.8.2. The boundary of subdomain B, line b, is interior to subdomain A, and the boundary of subdomain A, line a, is interior to subdomain B. The procedure is to guess the solution along one of the subdomain boundaries, say a. Using that $\Psi^1(a)$ as a temporary pseudo-boundary condition (Dirichlet), solve the interior problem in subdomain A using a direct method (herein, the marching method). This gives the solution everywhere in A, including $\Psi^1(b)$. Using this as the temporary pseudo-boundary condition (Dirichlet) for B, solve the interior problem in subdomain B using a direct method, giving $\Psi^2(a)$. Continue the iteration back and forth between the two subdomains until some iterative convergence criterion is met, e.g., $|\Psi^{n+1}(a) - \Psi^n(a)| < e$.

Explicit evaluation of residuals is not required, although this can be used in a rational iterative convergence criterion. The rate of convergence depends (dramatically) on the amount of overlap, and, of course, the iteration can be relaxed, extrapolated, etc. with the optimum relaxation factor r dependent on geometry and governing equations. (Analyses are useful [44,45] but difficult for general problems, and one should expect some numerical experimentation.) Significant overlap not only improves the convergence rate, it also drives the optimum relaxation factor r towards unity and makes the convergence history less sensitive to r. For reasonable overlap, it is possible to achieve convergence to e = 1.e-06 for a Poisson equation in 3 complete iterations (involving both subdomain solutions). (See Ehrlich, [4].) These would require only 2 subdomain initializations of a marching method (with small overlap, this is *less* operation count thanthe initialization on the entire domain), or just one subdomain initialization if sub-domains A and B, and their *types* of boundary conditions, were identical.

The amount of overlap is less important if a variation is used in which Dirichlet pseudo-boundary conditions are used on subdomain A, and gradient (Neumann) conditions are used for pseudo-boundary conditions on subdomain B; see Funaro et al. [46] and Widlund [47]. Mixed (Robin) conditions are also effective; see Marini and Quarteroni [48].

The original two-subdomain Schwarz method is now called the "multiplicative" Schwarz method [13]. Although algorithmically effective, it is *not* a parallel algorithm, but is essentially sequential; first one subdomain is solved with pseudo-boundary conditions, then the information is transferred to the pseudo-boundary conditions for the other subdomain. In another terminology common in point-iterative methods [12], it uses "immediate updating." In order to parallelize the method, one makes initial guesses not just for $\Psi^1(a)$ but also for $\Psi^1(b)$. Now the direct solution on A and B can proceed in parallel on multiprocessors, given $\Psi^2(a)$ and $\Psi^2(b)$ independently. Then the (low band width) boundary information transferred, and solutions are repeated for the second iteration. This is now called the "additive" Schwarz method [13].

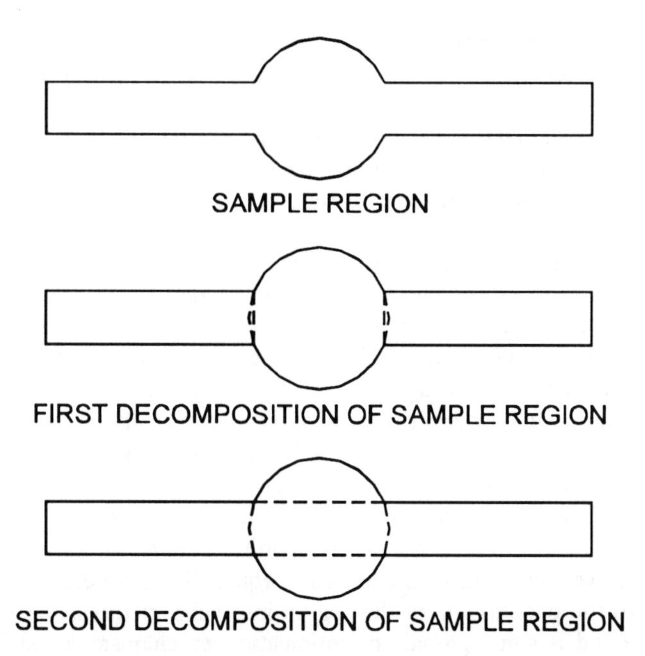

Figure 3.8.2. Domain Decomposition by Schwarz Alternating Procedure. (After Ehrlich [4] .)

Unfortunately, these two algorithms are analogous to the two oldest point-relaxation methods for the Poisson equation, Jacobi (also known as Richardson) iteration and Gauss-Seidel (also known as Liebman) iteration. The former is a natural parallel algorithm, while the latter is essentially serial. By choosing the algorithm naturally suited for a parallel machine, the analyst gives up a factor of 2 in convergence speed [12]. The same factor applies to multiplicative (serial) vs. additive (parallel) Schwarz for asymptotically small overlap (Bjorstad [13]) so that the algorithm achieves almost nothing, i.e., asymptotically, the use of two processors (assuming perfect parallel efficiency) does not compensate for the loss of algorithmic efficiency. This is an example of the trade-off (too commonly unacknowledged) between *algorithmic efficiency* and *machine efficiency* on parallel architectures. As Gropp and Keyes [10] note, "Perhaps the most subtle cost [of parallel computation] lies in algorithmic changes to 'improve' parallelism; by choosing a poor algorithm with better parallelism than another, less parallel algorithm, artificially good results can be found."

However, the situation improves for many subdomains. The penalty in algorithmic efficiency stays roughly at two, but the parallelism continues to improve, provided that alternating subdomains are updated together, e.g., alternating strip subdomains or red-black ordering. (Only half of the problem can be processed in parallel at one time.) The Dirichlet-Neumann pseudo-boundary conditions can also be extended to many subdomains, in various patterns (Camarero and Quarteroni [46]) and the application to higher dimensions is also straightforward. The additional coding is easy (once the subdomain marching solution is modularized) and there is no requirement for the overlapped regions to share the same node point locations (or even the same governing equations) as long as information is consistently passed between subdomains (e.g., by interpolation or least-squares weighting). In this most general form, the approach is sometimes known in the aerodynamics community as the Chimera scheme, pioneered by Steger [49]. Application within intra-time-step iterations is natural, and will often have the usual advantage of a good initial guess for the pseudo-boundary conditions. Steger [49] even allowed for time-dependent overlap to accommodate moving boundaries, e.g., the stores separation problem of applied aerodynamics.

Although the importance of the Schwarz alternating procedure is sometimes minimized in the mathematics community (perhaps because it is so straightforward and intuitive), it is in fact a very effective procedure. To quote P. L. Lions [50]: "In some sense, even if many interesting and important variants have been introduced recently, the Schwarz algorithm remains the prototype of such methods and also presents some properties (like 'robustness', or indifference to the type of equations considered) which do not seem to be enjoyed by other methods."

The performance of this or any multiblock iteration scheme is fundamentally limited by the rate of intra-block information transfer. Consider a domain decomposition consisting of n slabs, each covering the entire domain in I direction. Application of the additive (parallel) Schwarz procedure allows information to pass from one block only to the next adjacent block in one iteration. For an error in the initial guess in the lowest block to communicate with the uppermost block, it will require (n-1) iterations, setting a lower limit on the number of iterations. (For more general subdomain structuring, the number of iterative steps must be at least equal the diameter of the dual graph corresponding to the partitioning of the region into substructures; see Widlund [47].)

What is needed is some global communication mechanism, as in line SOR versus point SOR iterative methods (e.g., see Roache [12]). For the case of many subdomains, especially with a structured domain decomposition, the effective methods often take on the flavor of a nodal-iterative methods defined on a macromesh, with the now familiar variants of multigrid methods, preconditioned conjugate gradient, general preconditioning, Lanczos, multicolor ordering, etc. Alternately, instead of building up the hierarchy from node-equations to a macromesh, it can be conceptualized as starting from the larger "elements" and building down in "substructuring". See, e.g., Pasciak [40], Widlund [47], Glowinski and Wheeler [51], Nour-Omid et al. [52], Mandel and McCormick [53,54], Brochard [8]. In Widlund's [47] problem, effective multigrid-like methods can reduce the initial error by 1.e-04 in 7-10 global iterations. For additional work on Domain Decomposition in general, see Glowinski et al. [6], Chan et al. [7], Chan et al. [57], Glowinski et al. [58], Borgers and Widlund [68], Dryja and Widlund [69], Le Tallec [70], Keyes et al. [59], Keyes and Gropp [71], Keyes and Gropp [72], and Keyes [73-76].

Preliminary experiments by J. D. Morris and D. E. Keyes [62] indicate that simple marching methods can be quite competitive in Domain Decomposition as preconditioners for a Krylov-Schwarz algorithm.

3.9 Higher Precision Arithmetic; Applications on Workstations and Virtual Parallel Networks

The stability characteristics of the simple and extended marching methods described in Chapters 1-3 depend on the word-length of the computer used, or better, on the precision of the floating point arithmetic. Thus, an obvious way of extending the mesh size is to use higher precision. The cost is machine dependent. For some computers, double precision simply doubles the execution time; for others, the penalty is more or less severe. The cost of *serial* operations may more than double; furthermore, on vector machines (notably Cray computers) the use of double precision effectively stops the pipeline processing, eliminating the gains of this architecture. Still higher precision (e.g., quadruple precision) is typically very expensive. A *variable precision* computer (and language) would allow arbitrary sizing of the problem, but probably at large expense. The use of higher precision arithmetic in its most straightforward application combines with the previously described methods in a simple and obvious way.

Applications of the marching methods on minicomputers and super-microcomputers are different primarily because of these word length considerations. On the VAX (and similar) computers the precision for floating point operations is only about 8 significant decimal figures in single precision (SP) arithmetic. Some applications are possible, but the stringent requirements on word length caused by the marching instability (Chapter 1) make these generally impractical except for small problems with large cell aspect ratios.

Double precision (DP) calculations are clearly desirable for these machines. On some computers, advantage can be taken of the fact that the *marching algorithm does not require a completely DP calculation*. The stencil coefficients $C1$ - $C10$ do not need to be in DP, nor does dependent variable F. The march itself is performed in DP using only a few one-dimensional work arrays. For example, to represent $F(I,J)$ near the J-line of the march, we use $E1(I)$ for $F(I,J)$, $E2(I)$ for $F(I,J+1)$, etc., where $E1$, $E2$, etc. are DP. The algorithm is applied to the march of $E1$, $E2$ without the DP answer being actualized in two dimensions. The SP answer is stored in $F(I,J)$ during the last sweep with a DP to SP replacement statement like $F(I,J+1) = E2(I)$. The marching equation itself thus involves mixed arithmetic of SP arrays $C1$ - $C10$ and DP vectors $E1$, $E2$, etc. For the 9-point operator, the tridiagonal solution subprogram (and where applicable, the periodic tridiagonal solution subprogram) must also be done in DP. The only two-dimensional DP declaration required for the "basic" (i.e., un-stabilized) code is for the influence coefficient array CI. The Gaussian elimination subprograms must also be DP (we again use the LINPACK [58] or the later LAPACK [59] routines) and require some additional but one-dimensional DP work arrays.

Timing tests were run on a VAX 8810 with floating point hardware running under VMS Fortran 77. The penalty of operating a code entirely in DP is about 60% for optimized codes, and in the present case the penalty is less than 50%. Furthermore, DP on a VAX gives about 16 decimal figures of precision, compared to about 14 for SP on the Cray machines, so the stability is actually enhanced on the smaller computer.

Overall, considering the advantages of virtual memory for the VAX and double precision arithmetic on only one two-dimensional array, the marching methods are seen to be excellently suited for small computers. On any serial (i.e., non-vector) machine that does not penalize double-precision arithmetic by more than a factor of 2 in computation speed, the simplest and quite effective method of extending the mesh size is to use DP arithmetic selectively. (On some workstations, all arithmetic is performed in DP by default, with SP specification requiring truncation, so that DP costs less than nothing in speed, but still requires more memory.) This DP arithmetic can also be used in conjunction with the other methods for extending the mesh size.

The considerations above for workstations and high-end PC's are especially relevant to *virtual parallel networks*, in which heterogeneous nodes on a computer network are utilized as available via software such as PVM [60] and P4 [61]. This highly significant development in what might be called a "poor man's parallel computer" has the same appeal relative to parallel supercomputers as individual PC's and workstations have relative to serial supercomputers: *autonomy* !

References for Chapter 3

1. P. J. Roache, Marching Methods For Elliptic Problems: Part 1, *Numerical Heat Transfer*, Vol. 1, 1978, pp. 1-25. Part 2, *Numerical Heat Transfer,* Vol. 1, 1978, pp. 163-181. Part 3, *Numerical Heat Transfer*, Vol. 1, 1978, pp. 183-201.
2. H. A. Schwarz, Uber einege Abbildungsaufgaben, *Gesammelte Mathematische Abhandlungen*, Vol. 11, 1869, pp. 65-83.
3. H. A. Schwarz, *Gesammelte Mathematische Abhandlungen*, Springer, Berlin, Germany, Vol. 2, 1890, pp. 133-134.
4. L. W. Ehrlich, The Numerical Schwarz Alternating Procedure and SOR, *SIAM Journal of Scientific and Statistical Computing*, Vol. 7, No. 3, pp. 989-993.
5. K. Miller, Numerical Analogs to the Schwartz Alternating Procedure, *Numerical Mathematics*, Vol. 7, 1965, pp. 91-103.
6. R. Glowinski, G. Golub, G. A. Meurant, and J. Periaux, *First International Symposium on Domain Decomposition Methods for Partial Differential Equations*, Ecole Nationale des ponts et Chaussees, Paris, France, Jan. 7-9, 1987, SIAM, Philadelphia, 1988.
7. T. F. Chan, R. Glowinski, J. Periaux, and O. B. Widlund, (eds.), Domain Decomposition Methods, *Proc. Second International Symposium on Domain Decomposition Methods*, Los Angeles, CA, January 14-16, 1988, SIAM, Philadelphia, 1989.
8. L. Brochard, Efficiency of Multicolor Domain Decomposition on Distributed Memory Systems, in T. F. Chan et al. (Ref. 7), 1989, pp. 249-259.
9. M. Haghoo and W. Proskurowski, Parallel Efficiency of a Domain Decomposition Methods, in T. F. Chan et al. (Ref. 7), 1989, pp. 269-281.
10. W. D. Gropp and D. E. Keyes, Domain Decomposition on Parallel Computers, in T. F. Chan et al. (Ref. 7), 1989, pp. 260-268.
11. R. Lohner and K. Morgan, Domain Decomposition for the Simulation of Transient Problems in CFD, in R. Glowinski et al. (Ref. 6), 1988, pp. 426-431.
12. P. J. Roache, *Computational Fluid Dynamics*, rev. printing, Hermosa Publishers, Albuquerque, NM, 1976.
13. P. E. Bjorstad, Multiplicative and Additive Schwartz Methods: Convergence in the Two-Domain Case, in T. F. Chan et al. (Ref. 7), 1989, pp. 147-159.
14. I. Hirota, T. Tokioka, and M. Nishiguchi, A Direct Solution of Poisson's Equation by Generalized Sweep-Out Method, *Journal of the Meteorological Society of Japan*, Ser. II, Vol. 48, No. 2, Apr. 1970, pp. 161-167.
15. H. B. Keller, *Numerical Methods for Two-Point Boundary Value Problems,* Blaisdell, Waltham, MA, 1968, pp. 61-68.
16. D. E. Dietrich, Development of Block Implicit Relaxation (BIR) for Application to Semi-Implicit Ocean Models, Naval Oceanographic Laboratory Final Report, Contract No. N00014-77-C-0208, 1978.
17. D. Dietrich, B. E. McDonald, and A. Warn-Varnus, Optimized Block-Implicit Relaxation, *Journal of Compuational Physics*, Vol. 18, No. 4, 1975, pp. 421-439.
18. J. W. Poukey, J. R. Freeman, and G. Yonas, Simulation of Relativistic Electron Beam Diodes, *Journal of Vaccuum Science Technology*,Vol. 10, No. 6, 1973, pp. 954-958.
19. B. L. Buzbee, F. W. Dorr, J. A. George, and G. H. Golub, The Direct Solution of the Discrete Poisson Equation on Irregular Regions, Computer Science Dept. Technical Report, CS-71-195, Stanford University, Stanford, CA, 1970. See also *SIAM Journal of Numerical Analysis*, Vol. 8, 1971, p. 722.

20. R. W. Cottle, Manifestations of the Schur Complement, *Linear Algebra with Applications*, Vol. 8, 1974, pp. 189-211.
21. T. F. Chan and D. C. Renasco, A Framework for the Analysis and Construction of Domain Decomposition Preconditioners, in R. Glowinski et al. (Ref. 6), 1988, pp. 217-230.
22. A. J. Russo, Renormalization Methods for Several Marching Algorithms Used in The Direct Solution of Elliptic Equations, Sandia National Laboratories Report, SAND-78-0736, Albuquerque, NM, Apr. 1978.
23. G. Dahlquist and Å. Björk, *Numerical Methods*, Translation N. Anderson, Prentice-Hall, Englewood Cliffs, NJ, 1974.
24. J. W. Lucey and K. F. Hansen, A Stable Method of Matrix Factorization, *Transactions of the American Nuclear Society*, Vol. 7, 1964, p. 259.
25. D. R. Edwards and K. F. Hansen, The Stabilized March Technique Applied to the Diffusion Equation, *Nuclear Science Engineering*, Vol. 25, 1966, pp. 58-65.
26. R. Coleman, The Numerical Solution of Linear Elliptic Equations, *Journal of Lubrication Technology, Transactions ASME*, Ser. F, Vol. 90, 1968, pp. 773-776.
27. R. E. Bank and D. J. Rose, An $O(n^2)$ Method for Solving Constant Coefficient Boundary Value Problems in Two Dimensions, *SIAM Journal of Numerical Analysis*, Vol. 12, No. 4, 1975, pp. 529-539.
28. R. E. Bank and D. J. Rose, Marching Algorithms for Elliptic Boundary Value Problems I. The Constant Coefficient Case, *SIAM Journal Numerical Analysis*, Vol. 14, No. 5, 1977, pp. 792-829. (See also Aiken Computation Laboratory, TR 14-75, Harvard University, Cambridge, MA, 1975.)
29. R. E. Bank, Marching Algorithms for Elliptic Boundary Value Problems II. The Variable Coefficient Case, *SIAM Journal Numerical Analysis*, Vol. 14, No. 5, 1977, pp. 952-970. (See also Aiken Computation Lab. TR 16-75, Harvard University, Cambridge, MA, 1975, and R. E. Bank, Marching Algorithms and Block Gaussian Elimination, *Proc. Argonne Conference on Sparse Matrix Computations*, Argonne, IL, 1976.
30. M. E. Scott and H. A. Watts, Computational Solutions of Linear Two-Point Boundary Value Problems Via Orthonormalization, *SIAM Journal of Numerical Analysis*, Vol. 14, No. 1, 1977, pp. 40-90.
31. R. Madala, An Efficient Direct Solver for Separable and Non-Separable Elliptic Equations, *Monthly Weather Review*, Vol. 106, No. 12, Dec. 1978, pp. 1735-1741.
32. S. Y. K. Yee, An Efficient Method for a Finite difference Solution of the Poisson Equation on the Surface of a Sphere, *Journal of Computational Physics*, Vol. 22, No. 2, 1976, pp. 215-228.
33. C. Beckermann, C. Fan, and J. Milhailovic, Numerical Simulations of Double-Diffusive Convection in a Hele-Shaw Cell, *International Video Journal of Engineering Research*, Vol. 1, 1991, pp. 71-82.
34. M. R. Baer and R. J. Gross, A Two-Dimensional Flux-Corrected Solver for Convectively Domianted Flows, Sandia National Laboratories, Report No. 85-0613, 1985.
35. D. E. Dietrich, M. G. Marietta, and P. J. Roache, An Ocean Modeling System With Turbulent Boundary Layers and Topography: Numerical Description, *International Journal for Numerical Methods in Fluids*, Vol. 7, Sept. 1987, pp. 833-855.
36. D. E. Dietrich, P. J. Roache, and M. G. Marietta, Convergence Studies with the Sandia Ocean Modeling System, *International Journal for Numerical Methods in Fluids*, Vol. 11, 1990, pp. 127-150.
37. D. E. Dietrich, and P. J. Roache, An Accurate Low Dissipation Model of the Gulf of Mexico Circulation, *Proc. International Symposium on Environmental Hydraulics*,

University of Hong Kong, J. H. W. Lee, et al., (eds.), H. Balkeema, Amsterdam, December 14-16, 1991.

38. D. E. Dietrich, Sandia Ocean Modeling System Programmer's Guide and User's Manual, SAND92-7286, Sandia National Laboratories, Albuquerque, NM, Jan. 1993.
39. W. P. Tang, Wavefront Elimination and Renormalization, in T. F. Chan et al. (Ref. 7), 1989, pp. 235-246.
40. J. E. Pasciak, Domain Decomposition Preconditioners for Elliptic Problems in Two and Three Dimensions: First Approach, in R. Glowinski et al. (Ref. 6), 1988, pp. 62-72.
41. G. Meurant, Incomplete Domain Decomposition Preconditioners for Nonsymmetric Problems, in T. F. Chan et al. (Ref. 7), 1989, pp. 219-225.
42. P. Fischer, E. M. Ronquist, D. Dewey, and A. T. Patera, Spectral Element Methods: Algorithms and Architectures, in R. Glowinski et al. (Ref. 6), 1988, pp. 173-197.
43. A. A. Dorodnicyn, A Review of Methods for Solving the Navier-Stokes Equations, in H. Cabannes and R. Temam (eds.), *Lecture Notes in Physics, Vol. 18, Proc. Third International Conference on Numerical Methods in Fluid Mechanics*, Springer, New York, 1973, pp. 1-11.
44. D. J. Evans and K. Li-Shan, New Domain Decomposition Strategies for Elliptic Partial Differential Equations, in T. F. Chan et al. (Ref. 7), 1989, pp. 173-191.
45. K. Li-Shan, Domain Decomposition Method and Parallel Algorithms, in T. F. Chan et al. (Ref. 7), pp. 207-218.
46. D. Funaro, A. Quarteroni, and P. Zanolli, An Iterative Procedure with Interface Relaxation for Domain Decomposition Methods, *SIAM Journal of Numerical Analysis*, Vol. 25, No. 6, Dec. 1988, pp. 1213-1236.
47. O. B. Widlund, Iterative Substructuring Methods: Algorithms and Theory for Elliptic Problems in the Plane, in R. Glowinski et al. (Ref. 6), 1988, pp. 113-128.
48. L. D. Marini and A. Quateroni, An Iterative Procedure for Domain Decomposition Methods: A Finite Element Approach, in R. Glowinski et al. (Ref. 6), 1988, pp. 129-143.
49. J. L. Steger, F. C. Dougherty, and J. A. Benek, A Chimera Grid Scheme, ASME FED-Vol. 5, *Advances in Grid Generation, ASME Applied Mechancis, Bioengineering and Fluids Engineering Conference*, Houston, TX, June 20-22, 1983. K. N. Ghia and U. Ghia, (eds.), pp. 59-69.
50. P. L. Lions, On the Schwartz Alternating Method II: Stochastic Interpretation and Order Properties, in T. F. Chan et al. (Ref. 7), 1989, pp. 47-70.
51. R. Glowinski, and M. F. Wheeler, Domain Decomposition and Mixed Finite Element Methods for Elliptic Problems, in R. Glowinski et al. (Ref. 6), 1988, pp. 144-172.
52. B. Nour-Omid, B. N. Parlett, and A. Raefsky, Comparison of Lanczos with Conjugate Gradient Using Element Preconditioning, in R. Glowinski et al. (Ref. 6), 1988, pp. 250-260.
53. J. Mandel and S. McCormick, Iterative Solution of Elliptic Equations with Refinement: The Two Level Case, in T. F. Chan et al. (Ref. 7), 1989, pp. 81-92.
54. J. Mandel and S. McCormick, Iterative Solution of Elliptic Equations with Refinement: The Model Multi-Level Case, in T. F. Chan et al. (Ref. 7), 1989, pp. 93-102.
55. J. J. Dongarra, C. B. Moler, J. R. Bunch, and G. W. Stewart, *LINPACK Users' Guide*, Society for Industrial and Applied Mathematics, Philadelphia, 1979.
56. E. Anderson et al., *LAPACK Users' Guide*, Society for Industrial and Applied Mathematics, Philadelphia, 1992.

57. T. F. Chan, R. Glowinski, J. Periaux, and O. B. Widlund, *Third International Symposium on Domain Decomposition Methods for Partial Differential Equations*, Houston, TX, March 20-22, 1989. SIAM, Philadelphia, 1990.
58. R. Glowinski, Y. A. Kuznetsov, G. A. Meurant, J. Periaux, and O. B. Widlund, *Fourth International Symposium on Domain Decomposition Methods for Partial Differential Equations*, Moscow, May 21-25 1990, SIAM, Philadelphia, 1991.
59. D. E. Keyes, T. F. Chan, G. A. Meurant, J. S. Scroggs, and R. G. Voight, *Fifth International Symposium on Domain Decomposition Methods for Partial Differential Equations*, Norfolk, VA, May 6-8, 1991. SIAM, Philadelphia, 1992.
60. A. Geist, A. Beguelin, J. Dongarra, W. Jiang, R. Manchek, and V. Sunderam, *PVM 3 User's Guide and Reference Manual*, Oak Ridge National Laboratory, TN, ORNL:/TM-12187, May 1993.
61. R. Butler and E. Lusk, *User's Guide to the p4 Parallel Programming System*, Argonne National Laboratory, IL, ANL-92/17, Oct. 1992.
62. J. D. Morris and D. E. Keyes, Marching Preconditioners for Krylov-Schwarz Methods in the Solution of Transport Problems, Feb. 1994.
63. D. E. Dietrich, personal communication, 1994.
64. D. E. Dietrich and J. J. Wormeck, An Optimized Implicit Scheme for Compressible Reactive Gas Flow, *Numerical Heat Transfer*, Vol. 8, 1985, pp. 335-348.
65. D. E. Dietrich, Mechanics and Design of Centrifugal Swirl Nozzles, *Encyclopedia of Fluid Mechanics*, Chapter 14, Vol. 3, N. P. Cheremisinoff, (ed.), Syntax International, Singapore, 1985, pp. 365-370.
66. D. E. Dietrich, Mathematical Models of Mixing and Atomisation, *Atomisation and Spray Technology*, Vol. 3, 1987, pp. 291-307.
67. D. E. Dietrich, A Partial Theory of Atomization in Internal Mix Swirl Nozzles, *Aerosol Science and Technology*, Vol. 12, 1990, pp. 654-664.
68. C. Borgers and O. B. Widlund, A Domain Decomposition Laplace Solver for Internal Combustion Engine Modeling, *SIAM Journal of Scientific and Statistical Computing*, Vol. 10, No. 2, March 1989, pp. 211-226.
69. M. Dryja and O. B. Widlund, Domain Decomposition Algorithms with Small Overlap, *SIAM Journal of Scientific Computing*, Vol. 15, No. 30, May 1994, pp. 604-62.
70. P. Le Tallec, (ed.), *Domain Decomposition Methods in Computational Mechanics: Computational Mechanics Advances*, Vol. 1, No. 2, February 1994.
71. D. E. Keyes and W. D. Gropp, A Comparison of Domain Decomposition Techniques for Elliptic Partial Differential Equations and their Parallel Implementation, *SIAM Journal of Scientific and Statistical Computing*, Vol. 8, No. 2, March 1987, pp. s166-s202.
72. W. D. Gropp and D. E. Keyes, Domain Decomposition Methods in Computational Fluid Dynamics, *International Journal for Numerical Methods in Fluids*, Vol. 14, 1992, pp. 147-165.
73. D. E. Keyes, Domain Decomposition Methods for the Parallel Computation of Reacting Flows, ICASE Report No. 88-52, NASA-Langley Research Center, Hampton, VA, September 1988.
74. D. E. Keyes, Domain Decomposition: a Bridge between Nature and Parallel Computers, ICASE Report No. 92-44, NASA-Langley Research Center, Hampton, VA, September 1992.
75. D. E. Keyes, Y. Saad, and D. G. Truhlar, (eds.), *Domain-based Parallelism and Problem Decomposition Methods in Computational Science and Engineering*, Proc. Minnesota Supercomputer Institute, April 25-26, 1994, SIAM, Philadelphia, 1995.

76. D. E. Keyes and J. Xu, (eds.), *Domain Decomposition Methods, Proc. Seventh International Conference on Domain Decomposition Methods in Science and Engineering*, October 23-26, 1993. SIAM, Philadelphia, 1995.

Chapter 4

BANDED APPROXIMATIONS TO INFLUENCE MATRICES

4.1 Introduction

Algebraically, the basic marching algorithm described in Chapter 1 will produce the exact answer in a finite number of steps. As always, the practical distinction between exact methods and approximate (and perhaps iterative) methods is obscured in an actual computation with fixed word length computers, when only three or four significant figures in the final answer often are to be believed. In this chapter, we will consider making approximations to the influence coefficient matrices described in Chapters 1–3, approximating the full matrices by banded matrices, following [1,2]. For some conditions of practical interest, significant savings in operation count and storage can be achieved with these techniques.

4.2 Banded Approximation to C

The influence coefficient matrix C for the basic marching method is defined by Eq. (1.2.23). Generally, this matrix is full. (It has a bandwidth $W \cong 2J$, so it is full for $J > I/2$.) For the simple discretized Poisson equation, Eq. (1.2.28), the diagonal elements C_{ll} are largest, but C is not diagonally dominant in the sense of $C_{ll} \geq \Sigma_{l=1, l \neq k}^{L} |C_{lk}|$. However, true diagonal dominance does occur for a large cell aspect ratio $\beta \equiv \Delta x / \Delta y$. In the limit $\beta \rightarrow \infty$. C becomes purely diagonal, and the $L \times M$ two-dimensional problem degenerates to L independent one-dimensional problems, each M long, and governed by the one-dimensional error propagation characteristics described in Section 1.2.

This suggests approximating C with a banded matrix CB. The crudest level of approximation is the diagonal matrix. Although this approximation applies as $\beta \rightarrow \infty$, it brutalizes the problem; since the inverse of a diagonal matrix is diagonal, this approximation destroys the elliptic nature of the problem in the x direction. (This is not true for iterative solutions, e.g., the iterative patching technique suggested by Dorodnicyn [3], which is an early Domain Decomposition paper. The iterations are driven by one-dimensional equations, but the converged solution has an elliptic domain of influence in both directions.)

The crudest approximation that preserves the two-dimensional elliptic nature of the problem is the tridiagonal banded matrix approximation. The bandwidth $W = 3$ for 1 nonzero elements above and below the diagonal. Generally,

$$W = 1 + 2B \tag{4.2.1}$$

where $B = 0$ gives $W = 1$ (diagonal), $B = 1$ is tridiagonal, $B = 2$ is pentadiagonal, $B = 3$ is septadiagonal, etc. Note that the inverse of even a tridiagonal matrix is full, so that the entire domain of influence of the full C is mimicked. The first solution with the banded matrix gives an approximation to the initial error vector **E**, by which the $\Psi_{i,2}$'s are corrected. (See Section 1.2.2.) Repeated solutions give iterative corrections to $\Psi_{i,2}$, which, when they converge, give the same solutions as the full C. Convergence is strongly aided by a large cell aspect ratio β, fewer mesh points in the march direction J, increased band- width $W = 1 + 2B$, and increased number of corrective iterations r. The mesh size

I normal to the marching direction has a relatively weak effect, except for $I > 100$, which adversely affects the conditioning of C and CB (see Section 1.2.5). Table 4.2.1 presents a sampling of cases, giving FEMAX = largest error in $\Psi_{i,J}$ obtained in the march. (Since this can be reset to the correct boundary value, the maximum error at interior points is less than FEMAX.)

The effect of first-derivative advection terms can be significant. As noted previously in Chapter 1, Section 1.3.1, the condition of $\mathrm{Re}_{cx} = 2$ makes C upper triangular when $O(\Delta^2)$ three-point centered differences are used for first derivatives. At $\mathrm{Re}_{cx} = 2$, we find that the largest elements are not on the diagonal, but that the elements on the two bands above the diagonal are larger in magnitude than the diagonal elements, for $\beta = 3$. Still, for large β's, the banding approximation can be accurate. For $\beta = 5$ with $\mathrm{Re}_{cx} = 2$, a pentadiagonal approximation with 1 correction gives three-figure accuracy in the field of an 11×11 mesh. Even for $\mathrm{Re}_{cx} = 10$ with $\beta = 10$, a pentadiagonal approximation with 1 iterative correction, or a septadiagonal approximation with no correction, gives maximum errors of 5×10^{-4}.

I × J	β	B	r	FEMAX
11 × 11	3.	3	1	5.0E-04
11 × 11	3.	3	2	2.0E-05
11 × 11	5.	2	2	2.1E-14
21 × 21	5.	3	0	5.0E-04
21 × 21	5.	3	1	1.4E-07
21 × 21	5.	4	0	6.5E-07
21 × 21	5.	4	1	2.0E-12
31 × 31	5.	3	4	3.6E-03
31 × 31	5.	3	9	6.3E-06
51 × 51	5.	3		Unstable
51 × 51	10.	3	2	6.2E-05
51 × 51	10.	3	4	4.2E-08
101 × 51	10.	3	2	1.8E-04
101 × 51	10.	3	3	4.9E-06
101 × 51	10.	3	5	3.9E-09
101 × 51	10.	4	2	2.0E-09
101 × 101	10.	Full	3	1.8E-06
101 × 101	10.	5		Unstable
101 × 101	10.	7	3	2.1E-06

TABLE 4.2.1. Accuracy of Banded Approximation for the Poisson Equation in Two Dimensions. The bandwidth W = 1 + 2B, β is the cell aspect ratio Δx/Δy, r is the number of corrective iterations, FEMAX is the largest error in Ψi,J at the end of the march, and the notation 5.0E-04 means 5.0 × 10- 4. From Roache [2].

The banded approximations for C also apply for higher-order elliptic equations (Section 2.4) and for multiple marching methods for extending the mesh size (Section 3.3) provided that the elements of C are ordered as in Eq. (2.4.3).

To significantly reduce the operation count for initialization, we use a similar approximation in the generation of CB. If it suffices to retain $F_l \pm B$ as the only nonzero elements of CB, then the generation of CB can be consistently approximated by retaining only $e_{i \pm 2B, j}$ in the error propagation equation. Another approximation is possible in the generation and storage of CB. For the Poisson equation or other constant coefficient equations, for $\Delta x > \Delta y$ and $L \cong M$, it is true that $C_{lm} \cong C_{l+1, m+1}$ away from the corners of C, and this can be exploited in the banded approximation as indicated below.

4.3 Operation Count and Storage for Banded *CB*

To estimate the operation count for a band matrix of width $W = 1 + 2B$, we follow [4] (p. 166). If the full matrix has $n \times n$ elements, and if $B << n$, the leading term in the operation count θ for the initial elimination decreases from $n^3/3$ to $nB(B + 1)*$ and $nB(B + 1)\pm$ without pivoting. [With pivoting, the term is $nB(2B + 1)$, but experience shows that pivoting is often unnecessary.] For repeat solutions, θ decreases from n^2* and $n^2\pm$ to $n(2B + 1)$ without pivoting or to $n(3B + 1)$ with pivoting. By using these and the approximation in the previous section for the generation of CB, we obtain the following operation counts for the reference $2D$ Poisson equation.

The error propagation equation [Eq. (1.2.18)] is now applied to only $(4B + 1)M$ points in each march, rather than $\lambda L \times M$ points. Also, the full number of L marches is needed only for general variable coefficient problems. For constant coefficient problems or with coefficients that vary only in y, a banded march at i near the center of the mesh gives the same values of $\mathbf{F}_i$ as the marches at $i \pm 1$ with $\mathbf{F}$ shifted. Only $\gamma = (1 + 4B)$ marches are need without symmetry in CB, or $\gamma = (1 + 2B)$ marches with symmetry in CB. For general variable coefficients, $\gamma = L$. Then θ_2 of Eq. (1.2.26) is reduced by the banding approximation to

$$\theta_{2B} = [2\gamma(4B + 1)M]*, [3\gamma(4B + 1)M]\pm \tag{4.3.1}$$

(The gain from null calculations is negligible for $M >> B$.) The leading terms of θ_3 for initial solution of the banded CB without pivoting is reduced from θ_3 of Eq. (1.2.27) to

$$\theta_{3B} \cong LB(B + 1)*, LB(B + 1)\pm \tag{4.3.2}$$

The initialization procedure thus requires $(\theta_{2B} + \theta_{3B})$ operations, or

$$\begin{aligned}\theta_{\text{init},B} = {} & [2\gamma(4B + 1)M + LB(B + 1)]*, \\ & [3\gamma(4B + 1)M + LB(B + 1)]\pm\end{aligned} \tag{4.3.3}$$

A single repeat solution would still involve two marches of Eq. (1.2.17), with θ_4 given by Eq. (1.2.30), plus the solution for $e_{i,2}$ for which θ_5 is reduced without pivoting from Eq. (1.2.32) to

$$\theta_{5B} = L(2B + 1)*, L(2B + 1)\pm \tag{4.3.4}$$

However, each subsequent iteration requires only one march, giving $\frac{1}{2}\theta_4$ from Eq. (1.2.30) plus θ_{5B}. If r corrective iterations are required to achieve the desired accuracy in the banding approximation, then the operation count for repeat solutions is $(1 + \frac{1}{2}r)\theta_4$ plus $(1 + r)\theta_{5B}$ or

$$\theta_{\text{rep},B} \cong \left[\left(1+\frac{1}{2}r\right)4LM+(1+r)L(2B+1)\right]*\,,$$
$$\left[\left(1+\frac{1}{2}r\right)8LM+(1+r)L(2B+1)\right]\pm \tag{4.3.5}$$

Assuming $L = M$, and counting each * and $\pm$ as an operation, we obtain from Eqs. (4.3.3) and (4.3.4)

$$\theta_{\text{init},B} \cong 5(4B + 1)\gamma M + 2B(B + 1)M \tag{4.3.6}$$

$$\gamma = 1 + 4B \qquad \text{for } C \text{ asymmetric with constant coefficients} \tag{4.3.7a}$$

$$\gamma = 1 + 2B \qquad \text{for } C \text{ symmetric with constant coefficients} \tag{4.3.7b}$$

$$\gamma = L = M \qquad \text{for variable coefficients} \tag{4.3.7c}$$

$$\theta_{\text{rep},B} \cong \left(1 + \frac{1}{2}r\right)12M^2 + 2(1 + r)(2B + 1)M \tag{4.3.8}$$

It is noteworthy that this banded scheme will initialize the constant coefficient case in only $O(M)$ operations (although the coefficient is large). Considering the more general case with variable coefficients, restricting to the practical values $B \ll M$, and retaining only the highest order terms, we obtain

$$\theta_{\text{init},B} \cong 5(4B + 1)M^2 \tag{4.3.9a}$$

$$\theta_{\text{rep},B} \cong \left(1 + \frac{1}{2}r\right)12M^2 \tag{4.3.9b}$$

From these equations, it is seen that the marching method with the banded approximation gives an operation count for $L = M$ that is optimal, that is, $\propto M^2$, even for initializations. However, the required number of iterations r may increase with M, so that repeat solutions could be suboptimal. In semidirect solutions of nonlinear equations (Chapter 8), the banded approximation CB would tend to make the NOS method [5, 6] competitive with the Split NOS method [6, 7] in operation count; the NOS method is competitive with the Split NOS method [6, 7] in operation count; the NOS method is somewhat easier to program and requires less storage. Also, the banded approximation would tend to make Newton iteration feasible, replacing the simple Picard iteration of the NOS method.

For repeat solutions, the highest-order operation count is not significantly affected by the bandwidth W, provided it is reasonably small, but neither is the gain over the repeat solution with full C from Eq. (1.2.34b) very significant. The significant gains are in the operation count for initialization, as seen by comparing Eq. (4.3.6) with Eq. (1.2.34a), and in *storage*.

The full matrix C requires $L \times L$ storage, while the banded approximation would require only $W \times L$ storage for CB itself, which is over-written by the storage for the LU decomposition of CB; the latter storage is also $W \times L$ if pivoting is not required, or it is bounded by $3/2\, W \times L$ even if pivoting is required [8]. (In our applications, we use the banded matrix solvers written by C. B. Moler, which are available on the LINPACK [9] library. These solvers require $3/2\, W \times L$ storage. They also provide information on the condition number of the matrix.)

4.4 Intrinsic Storage: Data Compression for Massively Parallel Computers

The banded approximation has further significance for storage, on computers (or more importantly, for individual processors on massively parallel computers such as the 1024 processor CM-2) that have severe data storage limitations. It is possible to solve some problems with a program storage requirement *less* than the total number of node points in the mesh! That is, a marching method using the banded approximation can be utilized as a *data compression* algorithm.

Consider first the reference problem for the Laplace equation, that is, $\zeta_{i,j} = 0$ in Eq. (1.2.15). If the conditions are such as to allow an accurate marching solution to be obtained, then the only information required to preserve the solution $\Psi_{i,j}$ at all (i,j) is: (1) the boundary values along B1, B3, and B4 in Fig. 1.2.1; (2) the vector of values along $\Psi_{i,2}$; and (3) the marching algorithm corresponding to Eq. (1.2.17). The solution $\Psi_{i,j}$ at all (i,j) can always be regenerated, for an operation count of

$$\theta_{reg} = (2LM) *, (4LM) \pm \tag{4.4.1}$$

A simple method of marching this solution in an $IL \times JL$ mesh using a variable dimensioned only $IL \times 3$ is given below in a Fortran program segment with $S \equiv \Psi$. Here, ILM1 $= IL - 1$, AL $\equiv \alpha \equiv (\Delta y/\Delta x)^2$, ALF $= 2(1 + \alpha)$, and the boundary values in B3(JL) and B4(JL) are prestored. The marching equation is Eq. (1.2.17) with $\zeta_{i,j} = 0$.

```
        ⋮
      DØ 4 JP1 = 3, JL
      J = JP1 - 1
      S(I,2) = B3(J)
      S(IL,2) = B4(J)
      DØ 1 I = 2, ILM1
1     S(I,3) = ALF*S(I,2) - AL*(S(I+1,2) + S(I-1,2)) - S(I,1)
      WRITE 2, JP1, (S(I,3), I = 1, IL)
2     FØRMAT(...
      DØ 3 I = 2, ILM1
      S(I,1) = S(I,2)
3     S(I,2) = S(I,3)
4     CØNTINUE
        ⋮
```

The WRITE statement would produce lines of values of $\Psi_{i,j+1}$.

Actually, it is not even necessary to store three lines of S. The left-hand side of statement 1 could be replaced by S(I,2) in Fortran conventions, and the second statement in DØ-loop 3 could be dropped. However, the scheme as described is also applicable to nine-point implicit marches, such as those from cross derivatives (see Section 1.3.6).

If a banded approximation has been used for C (not necessarily for the generation of C, but for its solution) then we can *solve* and *intrinsically store* the entire solution for an $I \times J$ mesh in $O(I + J)$ points. For the Laplace equation, with a banded approximation CB of bandwidth W, the total storage can be of $O[(3 + 3/2\,W)I + 2J]$ for solution and intrinsic storage, or just $O(3I + 2J)$ for intrinsic storage along without CB. These numbers apply for the case of nontrivial boundary values.

Variable coefficient problems and nonhomogeneous problems will generally require $O(I \times J)$ storage locations for each coefficient, but not always. The coefficients may sometimes be generated by an algebraic rule, or may be separable in I and J, requiring only one-dimensional storage. A particularly interesting example occurs from the biharmonic equation $\nabla^4\Psi = 0$ with nonseparable boundary conditions, solved as two Poisson equations:

$$\begin{aligned} \nabla^2\Psi &= \zeta \\ \nabla^2\zeta &= 0 \end{aligned} \tag{4.4.2}$$

Now, if boundary values of both Ψ and ζ along B1, B3, and B4 of Fig. 1.2.1 are stored, plus the two vectors $\Psi_{i,2}$ and $\zeta_{i,2}$, the solution can again be regenerated by marching Ψ and ζ simultaneously, using only 2 or 3 lines of storage for each. With DYS $\equiv \Delta y^2$ and $Z \equiv \zeta$, we have from Eq. (1.2.17),

```
      DØ 1 I = 2, ILM1
      S(I,3) = DYS*Z(I,2) + ALF*S(I,2) - AL*(S(I+1,2) + S(I-1,2)) - S(I,1)
1     Z(I,3) = ALF*Z(I,2) - AL*(Z(I+1,2) + Z(I-1,2)) - Z(I,2)
```

For nonlinear problems in fluid dynamics-heat transfer solved by semidirect iterations (Chapter 8), it appears that at least one storage of size $I \times J$ is required during the solution. However, once obtained, even the steady-state *nonlinear* solutions can be intrinsically stored. Consider Ψ = stream function, ζ = vorticity, T = temperature, and velocity $\vec{V} = (u,v)$ using a decoupled energy equation.

$$\begin{aligned} \nabla^2\Psi &= \zeta \\ \nabla^2\zeta - u\mathrm{Re}\frac{\partial\zeta}{\partial x} - v\mathrm{Re}\frac{\partial\zeta}{\partial y} &= 0 \\ \nabla^2 T - u\mathrm{Re}\frac{\partial T}{\partial x} - v\mathrm{Re}\frac{\partial T}{\partial y} - \frac{E}{\mathrm{Re}}\Phi &= 0 \\ \Phi = 2\left[\frac{\partial u}{\partial x}\right]^2 + 2\left[\frac{\partial v}{\partial y}\right]^2 &+ \left[\frac{\partial v}{\partial x} + \frac{\partial u}{\partial y}\right]^2 \end{aligned} \tag{4.4.3}$$

where Φ is the dissipation term. (For example, see [10], p. 187.) With the boundary values and values along $(i, 2)$ stored as before, the line of values at $\Psi_{i,3}$ is marched out. This allows evaluation of velocities at $(i, 2)$ from

$$
\begin{aligned}
u_{i,j} &= \frac{\Psi_{i,j+1} - \Psi_{i,j-1}}{2\Delta y} \\
v_{i,j} &= \frac{-\left(\Psi_{i+1,j} - \Psi_{i-1,j}\right)}{2\Delta x}
\end{aligned}
\tag{4.4.4}
$$

Then the line $\zeta_{i,3}$ can be solved, allowing the march to continue to $\Psi_{i,4}$, etc. If the Eckert number $E = 0$, then the temperature T can be marched out with ζ. If $E \neq 0$, the march to $T_{i,3}$ requires $\Phi_{i,2}$, which requires $\partial u/\partial y|_{i,2}$ which requires $\Psi_{i,4}$. The march line for T can be lagged behind the march for Ψ, requiring only one extra line of storage. Note that u and v can be calculated locally at each i and j, so that they do not require even one-dimensional storage.

For the nonlinear solution by semidirect iteration (Chapter 8), the minimum one storage of size $I \times J$ can be stored on slow external storage, and brought in to high-speed memory in blocks as needed in the marching procedure.

This idea of intrinsic storage can also be used with banded approximations for three-dimensional problems (see Chapter 5).

4.5 Banded Approximation to $\hat{C}$

The influence coefficients $\hat{C}$ used to extend the field size to J with one or more patches (see Chapter 3, Section 3.4) may be similarly approximated by banded matrices $\hat{C}B$. These approximations behave even better than the banded approximations for C; even for $\Delta x = \Delta y$, we find that $B = 3$ is a quite adequate approximation for most applications. However, the gain in operation count is not very significant. The primary advantage is the reduction in storage penalty, reducing the storage for $\hat{C}$ from $L \times L$ to $3/2\,(W \times L)$ where $W = 1 + 2B$ for $\hat{C}B$.

A more significant gain in operation count can be realized in the generation of $\hat{C}$. Instead of solving L (repeat) marches to generate the full $\hat{C}$ and then truncating it to a banded approximation, the idea [2] is to make a banded approximation within the generation of $\hat{C}$. (Chan [11] describes the use of "boundary probing vectors".) In Section 3.4, to establish the matrix elements $\hat{C}_{l\,m}$, we solved the homogeneous equations L times with $e_{m1,j\,p} = 1$ and all other $e_{m,j\,p} = 0$ to establish the m_1th column of $\hat{C}$, repeating for $m_1 = 1, L$. That is,

$$
\begin{aligned}
\hat{E}_1 &= (1,0,0,0,0,0,0,\cdots 0) \\
\hat{E}_2 &= (0,1,0,0,0,\cdots 0) \\
&\vdots \\
E_{ml} &= (0,0,0,0,\cdots 0,1,0,\cdots 0)
\end{aligned}
$$

each requiring a homogeneous solution. Instead, consider the following $p = 5$ bandwidth probing vector with obvious generalization to bandwidth p other than 5.

$$\hat{E}_1 = (1,0,0,0,0,1,0,0,0,0,1,\cdots)$$

$$\hat{E}_2 = (0,1,0,0,0,0,1,0,0,0,0,1,\cdots)$$

$$\hat{E}_3 = (0,0,1,0,0,0,0,1,0,0,0,0,1,\cdots)$$

$$\hat{E}_4 = (0,0,0,1,0,0,0,0,1,0,0,0,0,1,\cdots)$$

$$\hat{E}_5 = (0,0,0,0,1,0,0,0,0,1,0,0,0,0,1,\cdots)$$

The banded approximation to $\hat{C}$ is then achieved in p homogeneous solutions, independent of L. The operation count for generation of $\hat{C}$ is not quite reduced by a factor p/L because p/L may not be an integer and because there are very few null operations in the march (that is, $\tilde{\lambda} = 1$ in Section 3.4). As some compensation, the coding is easier than accounting for $\tilde{\lambda} < 1$. The factor p/L more closely applies to line tridiagonal marches required for 9-point operators.

As we have seen, large cell aspect ratios will greatly improve the banding approximation (not noted in [11]).

Again, this approach is not limited to the use of Marching Methods for the direct solver. Schumann [12] also "truncated" the capacitance matrix $\hat{C}$ in another early example of what would now be called a "Domain Decomposition" paper. Mu and Rice [13] recently considered an approximation to the interface capacitance matrix (or Schur complement), viewing it as a matrix function $f(T)$ where T is a simple interface matrix, $T = 2I + K$ where K is the discrete one-dimensional Laplacian on the interface. They seek simple approximations (rational functions of low degree) to $f(T)$ such that it only needs to be evaluated at a few interpolating points. Their example calculation shows that, with this preconditioner, the PCG method converges in only four steps. They note that the approach is applicable to elliptic marching methods as well. However, as we noted at the beginning of this Section 4.5, the principal gain in efficiency when marching methods are used comes not from the solution of $\hat{C}$ but from its generation.

References for Chapter 4

1. P. J. Roache, Marching Methods For Elliptic Problems: Part 1, *Numerical Heat Transfer*, Vol. 1, 1978, pp. 1-25.
2. P. J. Roache, Marching Methods For Elliptic Problems: Part 2, *Numerical Heat Transfer,* Vol. 1, 1978, pp. 163-181.
3. A. A. Dorodnicyn, A Review of Methods for Solving the Navier-Stokes Equations, in H. Cabannes and R. Temam (eds.), *Lecture Notes in Physics, Vol. 18, Proc. Third International Conference on Numerical Methods in Fluid Mechanics*, Springer, New York, 1973, pp. 1-11.
4. G. Dahlquist and Å. Björk, *Numerical Methods*, translation by N. Anderson, Prentice-Hall, Englewood Cliffs, NJ, 1974.
5. P. J. Roache, Finite Difference Methods for the Steady-State Navier Stokes Equations, Sandia National Laboratories Report, SC-RR-72-0419, Albuquerque, NM,

Dec. 1972. (See also *Lecture Notes in Physics,* Vol. 18, Springer, New York, 1973, pp. 138-145.

6. P. J. Roache, The LAD, NOS and Split NOS Methods for the Steady-State Navier-Stokes Equations, *Computers and Fluids*, Vol. 3, 1975, pp. 179-195.
7. P. J. Roache, The Split NOS and BID Methods for the Steady-State Navier-Stokes Equations, in R. D. Richtmyer (ed.), *Lecture Notes in Physics, Vol. 35, Proc. Fourth International Conference on Numerical Methods in Fluid Dynamics*, Springer, New York, 1975, pp. 347-352.
8. G. E. Forsythe, M. A. Malcolm, and C. B. Moler, *Computer Methods for Mathematical Computations*, Prentice-Hall, Englewood Cliffs, NJ, 1977, p. 56.
9. J. J. Dongarra, C. B. Moler, J. R. Bunch, and G. W. Stewart, *LINPACK Users' Guide*, SIAM, Philadelphia, 1979.
10. P. J. Roache, *Computational Fluid Dynamics*, rev. printing, Hermosa Publishers, Albuquerque, NM, 1976.
11. T. F. Chan, Boundary Probe Domain Decomposition Preconditioners for Fourth Order Problems, in T. F. Chan, R. Glowinski, J. Periaux, and O. B. Widlund, (eds.), *Domain Decomposition Methods, Proc. Second International Symposium on Domain Decomposition Methods*, Los Angeles, CA, January 14-16, 1988. SIAM, Philadelphia, 1989, pp. 160-167.
12. U. Schumann, Fast Elliptic Solvers and Three-Dimensional Fluid-Structure Interactions in a Pressurized Water Reactor, *Journal of Computational Physics*, Vol. 36, 1980, pp. 93-127.
13. M. Mu/ and J. R. Rice, Preconditioning for Domain Decomposition through Function Approximations, *SIAM Journal of Scientific Computing*, Vol. 15, No. 6, Nov. 1994, pp. 1452-1466.

Chapter 5

MARCHING METHODS IN 3D

5.1 Introduction

In this chapter, three successful approaches [1,2] to solving three-dimensional elliptic equations will be presented. First, a straightforward extension of the *2D* method is given. This will be seen to have a very bad operation count, which, however, can be overcome with banded approximations to the influence coefficient matrix if the cell aspect ratios are favorable. Then, an alternative approach is presented, using a discrete Fourier transform in the third direction, that is very successful for a limited class of problems. Both can be used successfully within *3D* Domain Decomposition. Finally, N-plane relaxation using marching methods can be applied within Domain Decomposition or multigrid methods.

5.2 Simple 3D Marching

For purposes of illustration, consider the *3D* Poisson equation on a rectangular domain:

$$\frac{\partial^2 \Psi}{\partial x^2} + \frac{\partial^2 \Psi}{\partial y^2} + \frac{\partial^2 \Psi}{\partial z^2} = \zeta(x,y,z) \tag{5.2.1}$$

Usual $O(\Delta^2)$ differences are used, with $\Delta z = Z/(k - 1)$ where K is the maximum index in the z direction, at maximum $z = Z$.

For the case shown in Fig. 5.2.1, the march proceeds along the y axis. The vector **E** is now composed of the internal elements on the plane $j = 2$, arranged in some order such as

$$E_m = \left(\Psi_{2,2,2}, \Psi_{3,2,2}, \Psi_{4,2,2}, \ldots, \Psi_{I-1,2,2}, \Psi_{2,2,3}, \Psi_{3,2,3}, \ldots, \Psi_{I-1,2,3}, \ldots, \right.$$
$$\left. \Psi_{2,2,K-1}, \Psi_{3,2,K-1}, \ldots, \Psi_{I-1,2,K-1}\right) \tag{5.2.2}$$

The vector **F** consists of the internal elements on the plane $j = (J - 1)$, taken in the same order. (The ordering will be important in the Section 5.6 on operation count for banded approximation in *3D*.)

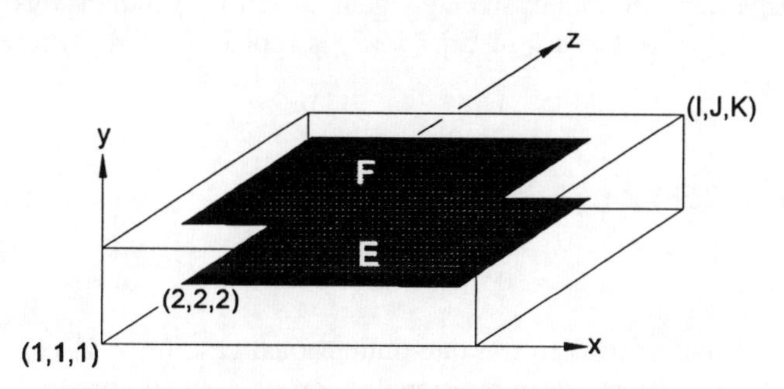

Figure 5.2.1. Notation for the 3-D EVP Marching Method.

Discretization of the Poisson equation gives the following replacement for the 2*D* march equation [Eq. (1.2.17)]:

$$\Psi'_{i,j+1,k} = \zeta_{ijk}\Delta y^2 + 2(1+\alpha_x+\alpha_z)\Psi'_{i,j,k} - \alpha_x(\Psi'_{i+1,j,k}+\Psi'_{i-1,j,k}) - \alpha_z(\Psi'_{i,j,k+1}+\Psi'_{i,j,k-1}) - \Psi'_{i,j-1,k} \tag{5.2.3}$$

where $\alpha_x = (\Delta y/\Delta x)^2$ and $\alpha_z = (\Delta y/\Delta z)^2$. All other coding aspects are the same as the 2*D* problem.

5.3 Error Propagation Characteristics for the 3*D* EVP Method

The error propagation equation for the 3*D* Poisson Eq. (5.2.3) is

$$e_{i,j+1,k} = 2(1+\alpha_x+\alpha_z)e_{i,j,k} - \alpha_x(e_{i+1,j,k}+e_{i-1,j,k}) - \alpha_z(e_{i,j,k+1}+e_{i,j,k-1}) - e_{i,j-1,k} \tag{5.3.1}$$

For both $\alpha_x \ll 1$ and $\alpha_z \ll 1$, this equation approaches the one-dimensional equation, which is merely linear in *j*. For $\alpha_x \gg \alpha_z$ or $\alpha_z \gg \alpha_x$, it approaches the two-dimensional propagation equation in *x* or *z* respectively, and the mesh limitation is given to good approximation by Eq. (1.2.38). For the case of $\alpha_x = \alpha_z = \alpha$, the symmetry of the error vector march algorithm (and the approximate independence of center values of *e* from the possible asymmetric boundary conditions along boundaries adjacent to the march) allows Eq. (5.3.1) to be written, only for the purpose of determining the error order of magnitude, as

$$e_{i,j+1,k} \cong 2(1+\alpha_{3D})e_{i,j,k} - \alpha_{3D}(e_{i+1,j,k}+e_{i-1,j,k}) - e_{i,j-1,k} \tag{5.3.2}$$

where $\alpha_{3D} = 2\alpha$. Since symmetry has eliminated the $(k \pm 1)$ indices, this Eq. (5.3.2) is the same as the two-dimensional equation [Eq. (1.2.18)] with α_{3D} replacing α. Likewise, Eq. (1.2.38) may be used for the three-dimensional case, replacing $\beta = \alpha^{-1/2}$ by $\beta_{3D} = \alpha_{3D}^{-1/2} = (2\alpha)^{1/2}$ or

$$\beta_{3D} = 2^{-1/2}\beta \cong 0.707\beta \tag{5.3.3}$$

Just as a curiosity, this arithmetic is readily generalized to *n* dimensions. For all ratios $\Delta y/\Delta x = \Delta z/\Delta x = \cdots = \alpha$, the α_{3D} in Eq. (5.3.2) is replaced by α_{nD} where

$$\alpha_{nD} = (n-1)\alpha \tag{5.3.4}$$

and the β in Eq. (1.2.38) is replaced by β_{nD} where

$$\beta_{nD} = (n-1)^{-1/2}\beta \tag{5.3.5}$$

The degeneracy of these forms to the one-dimensional case ($n = 1$) is consistent.

The approximate effect of dimensionality on the maximum mesh size may be demonstrated succinctly for this case, if β is large. Given some computer limitation (i.e., some value of *P*), it is seen from Eq. (1.2.38) that roughly, for $2 \leq \beta \leq 10$, the maximum

$Y/\Delta x = (J-1)\Delta y/\Delta x = (J-1)/\beta$ is independent of β in two dimensions. Generalizing with Eq. (5.3.5) we obtain, with fixed Δx,

$$\frac{Y_{nD}}{Y_2} = \frac{1}{\sqrt{n-1}} \tag{5.3.6}$$

This equation gives the approximate decrease in maximum Y for an n-dimensional problem compared with a $2D$ problem, for large cell aspect ratio. Although the jump from one to two dimensions greatly deteriorates the field limitations from an error propagation merely linear in j, it is seen that higher dimensions have a rapidly diminishing adverse effect. Changing from 2 dimensions to 3 dimensions reduces Y_2 by only $1/\sqrt{2}$ and to 10 dimensions by ⅓.

5.4 Operation Count and Storage Penalty for the 3*D* EVP Method

As the dimensions increase, the problem of inverting the full matrix C becomes more severe from the standpoints of speed, accuracy, and core storage requirements. We have $L \times M \times N$ unknowns, where $L = (I-2)$, $M = (J-2)$, $N = (K-2)$. The following development closely parallels the $2D$ operation count in Section 1.2.4; however, it should be noted immediately that the $3D$ counts are less meaningful because the Fortran subscripting operations can be more time-consuming than the operations shown here, at least for marching per se. (Single subscripting is advised, except for certain computers.)

Consider the first operation count for the initialization. Each application of the error propagation equation [Eq. (5.3.1)] requires 3* and 5±. [The coefficient $2(1 + \alpha_x + \alpha_z)$ is prestored. If $\alpha_x = \alpha_z$, one less * results, and if $\alpha_x = \alpha_z = 1$, we get only 1*, 5±.] Applied to each of $L \times M \times N$ points, this gives an operation count of

$$\theta_{31} = (3LMN)*,(5LMN)\pm \tag{5.4.1}$$

To establish C, we need $L \times N$ of these marches, reduced by null calculations by a factor λ_3, giving

$$\theta_{32} = \left(3\lambda_3 L^2 M N^2\right)*, \left(5\lambda_3 L^2 M N^2\right)\pm \tag{5.4.2a}$$

$$\lambda_3 = \left[1 - \frac{1}{3}\frac{L}{M}\right]\left[1 - \frac{1}{3}\frac{N}{M}\right] \qquad L \le M,\ N \le M \tag{5.4.2b}$$

The solution of the $LN \times LN$ influence coefficient matrix C by LU decomposition requires [see Eq. (1.2.27) with $n = LN$]

$$\begin{aligned}\theta_{33} = &\left[\frac{1}{3}L^3N^3 + \frac{1}{2}L^2N^2\right]*,\\ &\left[\frac{1}{3}L^3N^3 + \frac{1}{2}L^2N^2\right]\pm, \left[\frac{1}{2}L^2N^2\right]\div\end{aligned} \tag{5.4.3}$$

Thus, the $3D$ initialization procedure requires $(\theta_{32} + \theta_{33})$ operations, or

$$\theta_{3,\text{init}} = \left[3\lambda_3 L^2MN^2 + \frac{1}{3}L^3N^3 + \frac{1}{2}L^2N^2\right]*,$$

$$\left[5\lambda_3 L^2MN^2 + \frac{1}{3}L^3N^3 + \frac{1}{2}L^2N^2\right]\pm, \left[\frac{1}{2}L^2N^2\right]\div \tag{5.4.4}$$

The operation count for repeat solutions involves two marches [of Eq. (5.2.3)] and the solution for $e_{i,2,k}$. Careful programming can again eliminate the multiplication $\Delta y^2 \zeta_{i,j,k}$, and each application of the march equation requires 3* and 6± (or one less * for $\alpha_x = \alpha_y$, and two less * for $\alpha_x = \alpha_y = 1$) at each of $L \times M \times N$ points. For two marches, this gives

$$\theta_{34} = (6LMN)*, (12LMN)\pm \tag{5.4.5}$$

With the LU decomposition previously initialized, solutions for $e_{i,2,k}$ require [see Eq. (1.2.31) with $n = LN$]

$$\theta_{35} = (L^2N^2)*, (L^2N^2)\pm \tag{5.4.6}$$

With lower-order terms neglected, repeat solutions require $(\theta_{34} + \theta_{35})$ operations, or

$$\theta_{3,\text{rep}} = (6LMN + L^2N^2)*, (12LMN + L^2N^2)\pm \tag{5.4.7}$$

If, for comparison purposes, we again assume $L = M = N$ (for which $\lambda_3 = \lambda^2 = 4/9$) and count each "operation", then Eqs. (5.4.4) and (5.4.7) reduce to

$$\theta_{3,\text{init}} = \frac{2}{3}M^6 + \frac{32}{9}M^5 + \frac{3}{2}M^4 \tag{5.4.8}$$

$$\theta_{3,\text{rep}} = 2M^4 + 18M^3 \tag{5.4.9}$$

Unlike the $2D$ situation expressed in Eq. (1.2.34) these operation counts are worse than optimal even for repeat solutions. The $O(M^4)$ operation count for repeat solutions may be competitive for $3D$ Fourier series methods that are $O(M^3 \ln^2 M)$, but the $O(M^6)$ count for the initialization is disastrous. Only calculations that are to be performed *many* times in the same mesh, such as geophysical forecasting problems, are candidates for this method. Also, the storage requirement for C is M^4, which is very restrictive. (A $16 \times 16 \times 16$ mesh requires "64K" = 65,536 words just to store C.) However, it is possible to obtain an optimal operation count and reasonable storage penalty in $3D$ by using banded approximations to C.

5.5 Banded Approximations in 3D

As discussed in Section 4.2 for $2D$ problems, large cell aspect ratios allow the influence coefficient C to be approximated by a banded matrix CB. This is because the primary influence propagates along the y coordinate, so that $C_{nn} > C_{lm}$ for $l \neq m$. For $\Delta y \ll \Delta x$ in $2D$, we can neglect small enough elements in C, giving an approximate CB of band width B, where $B = 1$ gives tridiagonal CB, etc.

The analogous cutoff in $3D$ gives a much larger matrix CB. In $2D$, the major effect of the initial error $\mathbf{E}_m$ propagates to $\mathbf{F}_m$; retaining only these terms would give the (inadequate) diagonal approximation to C, so at least the neighbors $\mathbf{F}_{m\pm1}$ are retained to give the tridiagonal banded approximation in 2*D*. However, in 3*D*, the "neighbors" of $\mathbf{F}_m$ include those adjacent in both the x and z directions. If **F** were doubly subscripted in the xz plane (see Fig. 5.2.1) the neighbors of $\mathbf{F}_{ik}$ would be denoted by $\mathbf{F}_{i\pm1,\,k\pm1}$. In the single-subscripting convention of Eq. (5.2.2), which is more appropriate for band width interpretations, the immediate neighbors to $\mathbf{F}_l$ along x are $\mathbf{F}_{l\pm1}$, while the immediate neighbors along y are at $\mathbf{F}_{l\pm L}$ where $L = (I - 2)$. This is illustrated in Fig. 5.5.1 for a segment of the case of $L = N = 10$. The location (57) is the diagonal element considered, and the dashed lines enclose the neighbors in a pattern analogous to the tridiagonal approximation in 2*D*. The neighbors of 57 along x are 56 and 58, and along y are 47 and 67. (Consideration of wider band widths near corners makes it clear that the diagonal neighbors, 46, 48, 66, and 68 should also be included although they will be smaller than the x and y neighbors. In any case, they increase the band width only slightly compared with L.)

Generally, the 3*D* analog of a 2*D* banded approximation of band width $W = 1 + 2B$ (where $B = 1$ for tridiagonal, $= 2$ for pentadiagonal, etc.) is obtained with a band width W of

$$W = 1 + 2B\,(L + 1) \tag{5.5.1}$$

For a mesh with $N < L$, a narrower band width (with N replacing L above) could be obtained by changing the indexing on E and F to increase first in z rather than x.

Note that the prescription Eq. (5.5.1) for the banding approximation also includes many points that are not neighbors and that could be set to zero analogous to the 2*D* case, giving a "striped matrix" with stripe width of $(1 + 2B)$. However, the Gaussian elimination does not preserve the zeros between the stripes [4, p. 167], so no gain in the operation count results when using general software routines. (Some gain could be achieved by special programming to eliminate some zero operations.) The storage requirement [5] for the LU decomposition of CB is $[3B(L + 1) + 1]LN \cong 3BM$ for $L = M = N$; this is reduced from the simple 3*D* EVP method by a factor of $\cong 3B/M$. If cross derivatives like $\partial^2\Psi/\partial x\partial y$ are present, the tridiagonal solution increases the required band width to

$$W = (1 + 2B)L - 1 \tag{5.5.2}$$

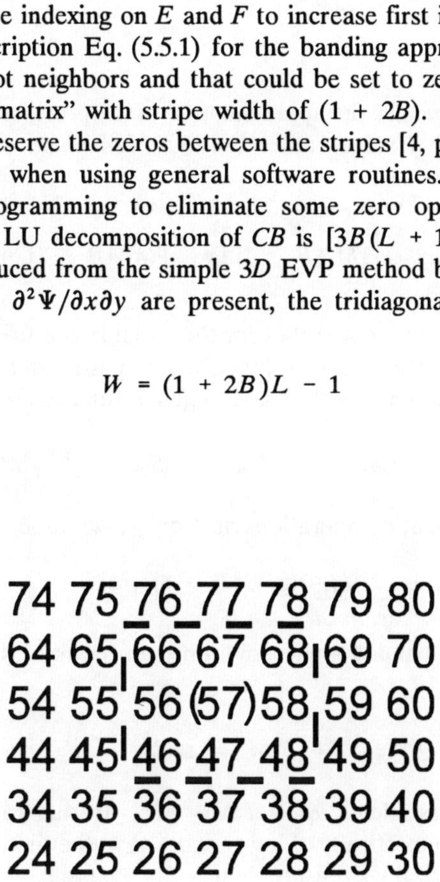

74 75 76 77 78 79 80
64 65 66 67 68 69 70
54 55 56 (57) 58 59 60
44 45 46 47 48 49 50
34 35 36 37 38 39 40
24 25 26 27 28 29 30

Figure 5.5.1. Banded Approximation in 3D for L = N = 10, B = 1

This banding approximation greatly reduces the operation count (next section) but still leaves us with an $O(M^5)$ contribution in the initialization because of the generation of CB. To reduce this, we use a similar approximation in the generation. In $2D$, if it suffices to retain $\mathbf{F}_{l\pm B}$ as the only nonzero elements of C_{lm}, then the generation can be consistently approximated by retaining only $e_{i\pm 2B,j}$ in the error propagation equation. In $3D$, the analogous approximation is to retain $e_{i\pm 2B,j,k\pm 2B}$. Then, in each march of the error propagation equation, at each j we calculate at only $(4B + 1)^2$ points instead of the entire plane of $L \times N$ points.

5.6 Operation Count for Banded Approximation in 3D

By using a banded approximation of width W given by Eq. (5.5.1), the $3D$ operation count is reduced as follows. According to [4, p. 166], for a band matrix with $n \times n$ elements in the full matrix where $p \equiv 1/2(W - 1)$ and $p \ll n$, the leading term in the operation count for the initial elimination decreases from $n^3/3$ to $np(p + 1)$ for * and $\pm$ without pivoting. [Experience shows pivoting is often unnecessary for these equations. If it is required, the corresponding term is approximately doubled, to $np(2p + 1)$.] In the notation of the $3D$ problem, $p = B(L + 1)$ and $L \gg 1$, so $p \cong p + 1 \cong BL$.

With these, the operation count for the initialization is estimated as follows. θ_{33} from Eq. (5.4.3) is reduced from a leading term of $1/3\, L^3N^3$ to $(LN)(BL)^2$. For $L = M = N$, this gives

$$\theta_{33B} \cong B^2M^4 \tag{5.6.1}$$

For the generation of C, we use the approximation of the previous section, calculating only $(4B + 1)^2$ points at each j value rather than the full plane of $L \times N$ points. Then Eq. (5.4.1) reduces to

$$\theta_{31B} = \left[3M(4B + 1)^2\right]*,\ \left[5M(4B + 1)^2\right]\pm \tag{5.6.2}$$

We still require $L \times N$ of these marches for the variable coefficient problem, but only γ^2 marches for the constant coefficient problem, where γ is given by Eq. (4.3.7). There are no significant null calculations for the banded generation, so Eq. (5.4.2a) reduces to

$$\theta_{32B} = \left[3(4B + 1)^2\gamma^2 M\right]*,\ \left[5(4B + 1)^2\gamma^2 M\right]\pm \tag{5.6.3}$$

For $L = M = N$, and counting operations as * or $\pm$, we obtain

$$\theta_{32B} \cong 8(4B + 1)^2\gamma^2 M \tag{5.6.4}$$

The initialization for the banded approximation then requires $(\theta_{33B} + \theta_{32B})$ operations, or

$$\theta_{3B,\text{init}} \cong B^2M^4 + 8(4B + 1)^2\gamma^2 M \tag{5.6.5}$$

For variable coefficient problems, $\gamma^2 \to LN = M^2$. In that case, although the second term of Eq. (5.6.5) is asymptotically of lower order than the first, it can be significant and even dominant for realistically sized problems. For example, with $B = 3$, the two terms given, $\theta_{3B,\text{init}} \cong (9M^4 + 1352M^3)$ and the latter term dominates for practical cases $(M < 150)$.

Similarly, we estimate the operation count for repeat solutions. From [4, p. 166], the full matrix operation count of M^2 (for * and $\pm$) decreases to $n(2p + 1)$ without pivoting [or about 50% more, to $n(3p + 1)$, with pivoting]. Then Eq. (5.4.6) changes to

$$\theta_{35B} \cong (2BL^2N)*, \quad (2BL^2N)\pm \tag{5.6.6}$$

Repeat solutions still involve two marches, with θ_{34} given by Eq. (5.4.5). However, each subsequent iteration requires only one march, giving $\frac{1}{2}\theta_{34}$, plus θ_{35B}. If r corrective iterations are required, the total θ is given by $(1 + \frac{1}{2}r)\theta_{34} + (1 + r)\theta_{35B}$, or

$$\theta_{3B,\mathrm{rep}} \cong \left[\left(1 + \frac{1}{2}r\right) 6LMN + (1 + r)2BL^2N\right]*,$$
$$\left[\left(1 + \frac{1}{2}r\right) 12LMN + (1 + r)2BL^2N\right]\pm \tag{5.6.7}$$

For $L = M = N$ and counting operations,

$$\theta_{3B,\mathrm{rep}} \cong \left[\left(1 + \frac{1}{2}r\right) 18 + (1 + r)4B\right] M^3 \tag{5.6.8}$$

Thus, the use of banded approximations allows an operation count for repeat solutions of the *3D* EVP method to remain "optimal." However, the problem size is limited both by storage and accuracy. The storage penalty is $[3B(L + 1) + 1]LN \cong 3BM^3$. A $21 \times 21 \times 21$ problem with $\beta = 10$ and $B = 2$ gives FEMAX $= 1.3 \times 10^{-6}$ in 2 corrective iterations; with $B = 3$, we get FEMAX $= 1.8 \times 10^{-10}$. Larger problems would have to be solved by an adaptation of the *2D* iteration scheme based on a direct solution on a submesh (see Section 3.7).

5.7 Additional Terms in the *3D* Marching Method

The *3D* marching method, like the *2D* method, is adaptable to variable coefficient terms, first derivatives, Helmholtz terms, cross derivatives, etc. The descriptions are fairly obvious carry-overs from the *2D* descriptions in Chapter 1, but the situation for cross derivatives in *3D* deserves some mention.

Assuming, as before, that the marching direction is y, the presence of terms like $\partial^2\Psi/\partial x\partial z$ does not change the explicit character of the march. If terms like $\partial^2\Psi/\partial x\partial y$ occur, the march is "one-dimensionally implicit," requiring the solution of N decoupled tridiagonal problems, each L long. Similarly, terms like $\partial^2\Psi/\partial y\partial z$ require a march with L decoupled tridiagonal problems, each N long. However, if both $\partial^2\Psi/\partial x\partial y$ and $\partial^2\Psi/\partial y\partial z$ are present, or if $\partial^3\Psi/\partial x\partial y\partial z$ is present, the march is "two-dimensionally implicit." The march in y proceeds one plane at a time, requiring a *2D* direct solution, by *2D* marching or another *2D* solver, at each step. (Conceivably, an iterative solution at the advance plane could be used if the equation there was strongly diagonally dominant.) Alternatively, some of the troublesome cross derivatives can be lagged and iterated, as Bank [6] does in *2D*.

The idea of intrinsic storage (see Chapter 4, Section 4.4) applies in *3D* as in *2D*. The methods for extending the mesh size (Chapter 3) can be used with banded approximations as above, but the coding is difficult and the storage penalties become serious.

5.8 3*D* EVP-FFT Method

As an alternative to the banded 3*D* marching method, the following method is available to treat 3*D* equations that are *separable* at least in the third direction.

Hockney's method in 2*D* [7,8] uses an FFT in one direction and, in the other direction, a one-dimensional matrix solution of the finite-difference equations for each Fourier component. To be applicable, the elliptic equation must be separable. The 3*D* EVP-FFT method presented here is similar. For a 3*D* elliptic equation and boundary conditions that are separable in the z direction, the dependent variable Ψ is transformed in z from $\Psi(x,y,z)$ to $\hat{\Psi}(x,y,k_z)$ where $\hat{\Psi}$ is complex and k_z is the z-wave number. Then the 2*D* equations, which need not be separable in x and y, are solved for each k_z component separately. Finally, the transformed variables $\hat{\Psi}$ are reassembled to give $\Psi(x,y,z)$.

Consider the 3*D* Poisson equation (5.2.1) with K mesh points and $N = (K - 2)$ unknowns in the z direction, and $NC = (K - 1)$ "cells." The application of a discrete Fourier transform to Eq. (5.2.1) gives

$$\frac{\partial^2 \hat{\Psi}}{\partial x^2} + \frac{\partial^2 \hat{\Psi}}{\partial y^2} = \hat{\zeta}(x,y,k_z) + \left[\frac{\pi k_z}{Z}\right]^2 \hat{\Psi} \tag{5.8.1}$$

This is a sequence of 2*D* complex Helmholtz equations, one for each of $NC/2$ values of k_z. The marching equations are

$$\hat{\Psi}_{i,j+1} = \Delta y^2 \hat{\zeta}_{i,j} + 2(1 + \alpha_x + \hat{\alpha}_z)\hat{\Psi}_{i,j} - \alpha_x(\hat{\Psi}_{i+1,j} - \hat{\Psi}_{i-1,j}) - \hat{\Psi}_{i,j-1} \tag{5.8.2}$$

where

$$\hat{\alpha}_z = \left[\frac{\pi k_z}{2N}\frac{\Delta y}{\Delta z}\right]^2 \tag{5.8.3}$$

5.9 Error Propagation Characteristics for 3*D* EVP-FFT Method

From the discussion of the 3*D* EVP error propagation characteristics above, it is clear that the EVP-FFT method will be most unstable for the highest wave numbers, and will be more unstable than the 3*D* EVP method using $O(\Delta^2)$ centered differences in z, because of the higher-order accuracy of the FFT. As in the case of the 1*D* Helmholtz equation discussed in Section 1.3.5, the method will be practical even without stabilization for $\pi \Delta y/\Delta z < 1$. With $\Delta x = \Delta z$, we find that the equivalent β for use in Eq. (1.2.38), the prediction of J_{max}, is reduced by about a factor of 0.4 compared with the 2*D* case. Thus, the demands on the cell aspect ratios $\Delta y/\Delta x$ and $\Delta y/\Delta z$ are more stringent than for 2*D* marching, but still practical for many problems with a preferred direction. Also the 2*D* stabilization methods (see Chapter 3) are readily applied to the 3*D* EVP-FFT method.

5.10 Operation Count and Storage Penalty for the 3*D* EVP-FFT Method

The original three-dimensional Poisson equation with NC cells in the third direction has been reduced to $NC/2$ two-dimensional complex Helmholtz equations. Additional work is involved in the FFT operations.

Each complex marching solution is equivalent to two real solutions, so there are equivalently NC two-dimensional real Helmholtz equations to be solved. However, there is a fortunate symmetry in the wave number k_z. For example, if $K = 17$ giving $NC = 16$, the 8 distinct wave numbers are

$$k_z = -4,\ -3,\ -2,\ -1,\ 0,\ +1,\ +2,\ +3 \tag{5.10.1}$$

Now the marching equation (5.8.1) is symmetric in k_z^2. (This will be true for any elliptic equation with only even-valued derivatives in z.) This means that the generation and solution of the influence coefficient matrix C for $k_z = -3$ is the same as C for $k_z = +3$, reducing this part of the operation count and storage requirement by $(NC/2 + 1)/NC \cong ½$ for NC large. In the following we will assume $NC \gg 1$ so $NC \cong N$. The storage required for the C's is then about $½ L^2N$.

The FFT is most efficient when the number of data points is a power of 2, so we assume $NC = (2^n - 1)$. Then the asymptotic operation count for the one-dimensional FFT over N data points [4, p. 414] is

$$\theta_{FFT} = (2N \ln N) \text{ complex } *, \quad (2N \ln N) \text{ complex } \pm \tag{5.10.2}$$

In [9], we have estimated that each of these complex operations requires roughly an equivalent time (on a scientific computer) of about 4.5 real operations, giving equivalently $9N \ln N$ real * and $\pm$. This count will apply to the backward transform to reassemble $\Psi(x,y,z)$ from $\hat{\Psi}(x,y,k_z)$ since $\hat{\Psi}$ is complex. However, for the forward transform on *real* ζ to obtain ζ for Eq. (5.8.2), a slightly more efficient packing procedure can be used within the FFT algorithm. For the range of N practical for discrete multidimensional solutions, we have estimated [9] a further decrease in θ_{FFT} by roughly a factor of ⅔, giving another $6N \ln N$ real * and $\pm$. The forward and backward FFTs are applied to λ of each of $L \times M$ lines of data, each N long, where λ is the reduction factor for 2D null calculations given by Eq. (1.2.26b). The total of the additional operations from the FFT calculation is thus

$$\theta_F = (15\lambda LMN \ln N) *, \quad (15\lambda LMN \ln N) \pm \tag{5.10.3}$$

The calculation of $\hat{\alpha}_z$ in Eq. (5.8.2) can be prestored like the other coefficients, so the operation count for the 2D Poisson equation also applies. These additional operations apply only to repeat solutions, since the initialization involves a homogeneous counterpart of Eq. (5.8.2), that is, there is no FFT operation on ζ, and no backward FFT on $\hat{\Psi}$ is required until a solution is calculated. The operation count for the initialization is just $N/2$ times the 2D value from Eq. (1.2.29), or

$$\begin{aligned} \theta_{\text{init,3D } EVP/FFT} &= \left[\lambda L^2MN + \frac{1}{6}L^3N + \frac{1}{4}L^2N\right] *, \\ &\left[\frac{3}{2}\lambda L^2MN + \frac{1}{6}L^3N + \frac{1}{4}M^2N\right] \pm, \quad \left[\frac{1}{4}L^2N\right] \div \end{aligned} \tag{5.10.4}$$

The repeat solutions require N times the 2D value from Eq. (1.2.33), plus the FFT operations from Eq. (5.10.3), or

$$\begin{aligned} \theta_{\text{rep,3D } EVP/FFT} &= \left(15\lambda LMN \ln N + 4LMN + L^2N\right) *, \\ &\left(15\lambda LMN \ln N + 8LMN + L^2N\right) \pm \end{aligned} \tag{5.10.5}$$

Counting each operation and again assuming $L = M = N$ so that $\lambda = ⅔$, we obtain

$$\theta_{\text{init,}3D\ EVP/FFT} = \frac{17}{9}M^4 + \frac{3}{4}M^3 \cong 2M^4 \tag{5.10.6a}$$

$$\theta_{\text{rep}3D\ EVP/FFT} = (20 \ln M + 14)M^3 \tag{5.10.6b}$$

Although not asymptotically optimal, the operation count for repeat solutions is $O(M^3 \ln M)$, which is better than $3D$ Fourier series methods of $O(M^3 \ln^2 M)$. Actually, for a practical range of M, Eq. (5.10.6) gives θ_{init} somewhat less than θ_{rep}.

The comparison of the $3D$ marching method using banded approximations (above) is different for initialization and repeat solutions, and depends on M and on B, the size of the banded approximations. As a basis of comparison, we consider M = 32, B = 3, and r = 1, and consider only the constant coefficient problem. Then comparison of Eq. (5.10.6a) and Eq. (5.6.5) shows the $3D$ EVP-FFT method to be about 4.5 times faster for initialization, and comparison of Eq. (5.10.6b) and Eq. (5.6.8) shows it to be about 1.6 times slower for repeat solutions. Storage considerations greatly favor the $3D$ EVP-FFT method, by a factor of $\sim 6B$ in the additional storage for C's. If out-of-core storage is required on a particular computer, such a storage advantage can result in a time advantage for large problems.

Most significantly, each Fourier component can be marched out on a *separate processor* in a parallel computer, a natural and load-balanced parallelism.

5.11 Accuracy and Additional Terms in $3D$ EVP-FFT Method

When the problem is periodic in z, the FFT is more accurate than $O(\Delta z^2)$ finite differences. (In some sense, it is "infinite order" [10].) But when the problem is not periodic in z, a "ringing" or Gibbs phenomena occurs, which slows truncation error convergence, especially near the boundaries. However, the method is still formally $O(\Delta^2)$ [11]. The "reduction-to-periodicity" technique described in [9] can improve the accuracy for nonperiodic problems, but does not appear to be applicable in a direct solution. Conceivably, it could find application within semidirect iterations for nonlinear equations (Chapter 8).

The presence of additional terms, like x and y advection terms or cross-derivative terms with variable coefficients, can readily be handled by the $3D$ EVP-FFT method so long as the third (z) coordinate is not involved. All coefficients must be functions of x and y only, and only constant coefficient second derivatives in z are allowed. (Terms like $\partial^4\Psi/\partial z^4$ are allowed, with a serious decrease in field size, provided that the higher-order boundary conditions are compatible with the FFT. These appear to be contrived problems, and for practical purposes we can say that the direct method is limited to only the term $\partial^2\Psi/\partial z^2$ in the z direction.) A spatially constant multiplier can, of course, be absorbed by scaling. Variable coefficients and other derivatives must be treated by iteration. A similar approach to a $3D$ Poisson solution using FFT coupled to a direct FACR algorithm was used by Reid et al. [12] but is even more restricted in the admissible terms.

The idea of intrinsic storage (Chapter 4, Section 4.4) applies to the $3D$ EVP-FFT method, as do the methods for extending the mesh size (Chapter 3). For the latter, slow external storage can be used for the extra matrices required, since each Fourier component is calculated separately.

5.12 *N*-Plane Relaxation Within Multigrid and Domain Decomposition Methods

Another use of marching methods in *3D* is by plane relaxation, or n-plane relaxation. Just as line relaxation (line SOR, ADI methods) uses a direct line solver (Thomas tridiagonal algorithm) in a line-by-line iteration to solve a *2D* or *3D* problem, so can we use a direct plane solver (*2D* marching methods, basic or stabilized) in a plane-by-plane iteration to solve a *3D* problem. The variations are immediately obvious: fixed $I \times J$ plane relaxation; alternating $I \times J$, $J \times K$, $K \times I$ planes; symmetric sequences of plane rotation; odd-even plane relaxation; under- or over-relaxation, etc. The direct plane solver can also be used in a semicoarsening multigrid, e.g. Schaffer [13].

For the model Poisson equation with equal grid increments, the improvement in speed is not significant, even for repeat solutions. For the classic multigrid problem, the performance in number of iterations is almost as good in *3D* as in *2D* (or even *1D*), the accuracy tolerance is adjustable (i.e., if you do not require high accuracy, you can iterate less, compared to the fixed cost of a direct solver), and the storage penalty is less than that of marching methods. So it is preferable to just use the multigrid methods without bothering with marching methods.

For more general problems, plane iteration using marching methods can be very useful, even critical. As discussed in Chapter 9, multigrid methods and marching methods are in many respects complimentary, and wherever multigrid methods have trouble is a good place to look for an application of marching methods. An important one is strong anisotropy of the coefficients. Multigrid methods can be successful, but may require a semicoarsening approach (e.g. Schaffer [13]), which is "multigrid" storage in only one direction and which significantly affects directionality, operation count, and storage penalty (see discussions in Chapter 9). A good candidate for planar semicoarsening multigrid (or other relaxation) is a problem with strong variation of coefficients and, perhaps, first derivative terms in (say) the $I \times J$ plane, solved by planar marching methods, and using 1D multigrid in Z.

This algorithm can also be conceptualized as a Domain Decomposition method wherein each subdomain is a plane. The distinction of methods is often fuzzy (as noted in the introductions to Chapters 1 and 9) and this is nowhere more evident than here. Multigrid methods can and have been re-interpreted as domain decomposition methods, as could line relaxation and ultimately point relaxation, with each node becoming a domain, but this is not a useful taxonomy. For a *2D* problem, n-line relaxation (usually called k-line, but here we reserve k for the third spatial index) involves direct solution of n lines at a time. For $n = 2$, this requires a bi-tridiagonal (or equivalently, a pentadiagonal) direct solver. Its iterative convergence rate is faster than ($n = 1$) relaxation, but of course, the operation count per iteration is higher. (For $n = J - 2$, all interior j-planes of the *2D* problem, we have an expensive *2D* direct solver). Since, for $n > 1$, we are directly solving on strips more than one node wide, n-line relaxation is fruitfully interpreted as a Domain Decomposition method. The usual variations of overlapping and non-overlapping domains apply, with $n = 3$ minimal for overlapping.

The analogous algorithm in *3D* is n-plane relaxation, again with overlapping and non-overlapping variations. Although applicable to serial and vector computers, the most attractive application is to parallel multiprocessor machines for large problems. From the viewpoint of optimizing the operation count and stability for the direct solutions on subdomains, we do *not* want "compact" subdomains, i.e., with $I = J = K$, but rather "thin" subdomains, e.g., $I \sim J \gg K$. I can be larger, limited by ill-conditioning to roughly $O(100)$ (see Chapter 1), but the operation count is improved somewhat for I smaller. J (the marching direction) will usually be smaller, limited by march stability to roughly $O(15)$ (or 30 for two-directional marching) for equal grid increments for the Poisson equation, but can be much larger for favorable cell aspect ratios. And K should be very small in order to minimize the operation counts and storage penalties for simple *3D* marching [see Eqs. (5.4.4) and (5.4.7), and Section 5.6 for banded approximations] or large for the *3D* EVP-FFT method. For simple marching, $K = 1$ (planar domain decomposition) would be optimal for the operation counts of the marching method, but would not be optimal overall since no overlap of domains is possible. Intuitively, it would be surprising if the optimum K did not allow for at least 3 fully interior planes and 1 overlap on each side, i.e., $K > 4$. Beyond that, intuition fails. The determination of near-optimal subdomain structuring for n-plane domain decomposition using *3D* marching methods is a challenge.

References for Chapter 5

1. P. J. Roache, Marching Methods For Elliptic Problems: Part 1, *Numerical Heat Transfer*, Vol. 1, 1978, pp. 1-25.
2. P. J. Roache, Marching Methods for Elliptic Problems: Part 2, *Numerical Heat Transfer,* Vol. 1, 1978, pp. 163-181. Part 3, *Numerical Heat Transfer,* Vol. 1, 1978, pp. 183-201.
3. A. A. Dorodnicyn, A Review of Methods for Solving the Navier-Stokes Equations, in H. Cabannes and R. Temam (eds.), *Lecture Notes in Physics, Vol. 18, Proc. Third International Conference on Numerical Methods in Fluid Mechanics*, Springer, New York, 1973, pp. 1-11.
4. G. Dahlquist and Å. Björk, *Numerical Methods*, translation N. Anderson, Prentice-Hall, Englewood Cliffs, NJ, 1974.
5. G. E. Forsythe, M. A. Malcolm, and C. B. Moler, *Computer Methods for Mathematical Computations*, 1977, Prentice-Hall, Englewood Cliffs, NJ, p. 56.
6. R. E. Bank, Marching Algorithms for Elliptic Boundary Value Problems II. The Variable Coefficient Case, *SIAM Journal of Numerical Analysis*, Vol. 14, No. 5, 1977, pp. 952-970.
7. R. W. Hockney, A Fast Direct Solution of Poisson's Equation Using Fourier Analysis, *Journal of the Association for Computing Machinery*, Vol. 12, No. 1, 1965, pp. 95-113.
8. R. W. Hockney, The Potential Calculation and Some Applications, in B. Alder, S. Fernbach, and M. Rotenberg (eds.), *Methods in Computational Physics, Vol. 9, Plasma Physics*, 1974, pp. 135-211.
9. P. J. Roache, A Pseudospectral FFT Technique for Non-Periodic Problems, *Journal of Computational Physics*, Vol. 27, No. 2, 1978, pp. 204-220.
10. S. A. Orszag and M. Israeli, Numerical Solution of Viscous Incompressible Flows, *Annual Review of Fluid Mechanics,* Vol. 6, 1974, pp. 281-318.
11. G. Sköllermo, A Fourier method for the Numerical Solution of Poisson Equation, *Mathematics of Computation*, Vol. 29, No. 131, July 1975, pp. 697-711.
12. D. Reid, A. Chan, and M. Al-Mudares, A Direct Solution of Poisson's Equation in a Three-Dimensional Field-Effect Transistor Structure, *Journal of Computational Physics*, Vol. 99, 1992, pp. 79-83.

13. S. Schaffer, An Efficient "Black Box" Semicoarsening Multigrid Algorithm for Two and Three Dimensional Symmetric Elliptic PDE's with Highly Varying Coefficients, *Proc. Fifth Copper Mountain Conference on Multigrid Methods*, March 31-April 5, 1991. Also, to appear in *SIAM Journal of Numerical Analysis*.

Chapter 6

PERFORMANCE OF THE 2D GEM CODE

6.1 Introduction

The present chapter describes timing and accuracy tests [1] on a particular realization of the marching methods, the GEM (General Elliptic Marching) codes. Timing and accuracy tests of the GEM codes are described. The GEM codes solve elliptic and mixed discretized two-dimensional partial differential equations by the direct (noniterative) spatial marching methods described in previous chapters and in [2]. Both 5-point and 9-point stencils may be solved, with no requirement that the coefficients be separable, and quite general boundary conditions are treated. The basic GEM codes depends on problem parameters (primarily a large cell aspect ratio $\Delta x/\Delta y$) to control the instability incurred in marching elliptic equations. For a 5-point operator with nonperiodic boundary conditions, repeat solutions are obtained in the time equivalent of two SOR iterations. Tests of both an earlier (unstabilized) and a later (stabilized) version are given. Stabilizing codes allow an increase of the problem size in the marching direction at some penalty in execution time and core storage.

6.2 Uses and Users

Elliptic and mixed equations with nonseparable coefficients arise in a variety of applications in heat transfer, fluid dynamics, electromagnetic field calculations, etc. Even the simple Poisson equation, one of the most common equations in mathematical physics, becomes nonseparable when written in general nonorthogonal coordinates. The simplest second-order finite-difference discretization then leads to a 9-point nonseparable stencil. Since excellent codes are already available for variables-separable elliptic problems (see especially the FISHPAK codes of Swarztrauber and Sweet [3]; see also [4] for a more specialized but high-order solution by the present author), it was decided to concentrate on the general nonseparable problem with the GEM codes. The GEM codes are notable for the generality of the problems that they can solve.

Even for the variables-separable problem, the present GEM codes are faster than other direct methods for repeat solutions. However, these codes (e.g., FISHPAK [3]) are more robust, since they are not sensitive to problem size or cell aspect ratio. It has been our observation that non-specialist users of the GEM codes can get quite upset when a problem change that seems insignificant to them destroys the accuracy of the solution. For a general user, we would recommend FISHPAK (available as TOMS algorithm 451; see [3]) for variables-separable problems and reserve the GEM codes for the nonseparable problems.

The intended audience of these codes is a user who has some knowledge and experience in the solution of discretized partial differential equations. The user is simply provided with documented FORTRAN subroutines and some example problems. In terms of Parlett's description [5], the GEM package would be at level 4 with codes like Nastran and finite-element packages, rather than the level 3 of LINPACK, EISPACK, etc. To quote Parlett's description, "Notice that the user with a problem is directly involved at this level. He manages the package but does not write the program." In the GEM codes, there is only minimal argument checking and not very extensive diagnostic printouts. Also, the finite-difference discretization is left to the user.

6.3 Overview of the GEM Codes

The GEM codes solve elliptic and mixed discretized two-dimensional partial differential equations by direct (noniterative) spatial marching methods. Both 5-point and 9-point stencils may be solved, with no requirement that the coefficients be separable. For example, GEM solves the usual second-order accurate discretization of

$$aF_{xx} + bF_{yy} + cF_x + dF_y + eF_{xy} + fF = g \tag{6.3.1}$$

where a, b, ..., g are all functions of x and y. The marching methods used in the GEM codes were described in detail in previous chapters. The basic GEM code is based on "simple marching" and depends on problem parameters to control the instability incurred in marching elliptic equations; for realistic physical problems, this primarily depends on a large cell aspect ratio $\beta = \Delta x/\Delta y$.

Operation counts θ were given in some detail in Chapter 1. For a simple Poisson equation (5-point operator) without making use of symmetry, in a square array, this gives

$$\theta_{\text{init}} = 4M^3 + 3/2M^2 \qquad \theta_{\text{rep}} = 14M^2 \tag{6.3.2}$$

The initiation (init) count is less than that required to establish a single solution by point SOR, and the repeat (rep) count is less than two point-SOR iterations. Since operation counts like these neglect many overhead and subscripting operations, it is necessary to validate them with actual timing tests, especially since Eq. (6.3.2) indicates such remarkable efficiency.

When a 9-point operator is used, the marching solution proceeds a line at a time (like line SOR) and requires a tridiagonal solution at $j + 1$ at each step in the march. This, of course, increases the operation counts, but not their order (i.e., repeat solutions are still optimal, with $\theta \propto M^2$). For other aspects of the method, see Chapters 1–5.

6.4 Problem Description in the Basic GEM Code

The code is written not with a "general user'" in mind, but for one who knows both finite differences and FORTRAN. The discretization of the continuum partial differential equation is left to the user. The code is written in FORTRAN, and the subroutine GEM solves the following stencil.

$$\begin{bmatrix} C_1 & C_2 & C_3 \\ C_4 & C_5 & C_6 \\ C_7 & C_8 & C_9 \end{bmatrix} F_{ij} = C_{10} \tag{6.4.1}$$

In the original code on which the presently reported tests were performed, all the coefficients C_1, C_2, ..., C_{10} were arrays stored in the labeled COMMON block GEMCOM. (The smart user could change some or all of these to BLANK COMMON for storage efficiency.) In a later FORTRAN 77 version, the $C1$–$C10$ are in argument lists, easily changed via INCLUDE statements. The Subroutine Call is of the following form.

```
CALL GEM (INIT, F, IL, JL, ILD,
        N59, IPER, ICOR, EMX, NRC, RCOND, JMAR,
        JBOT, JTOP, NDBC, FDBC,                              (6.4.2)
        IPVT, CI, KLD, NC10)
```

INIT = 0 initiates only, = 1 initiates and solves, > 1 backsolves only. The solution is stored in F. The problem size is IL × JL, with the actual first DIMENSION of the arrays being ILD. N59 = 5 or 9 gives the 5-point or 9-point operator solution. (If N59 = 5, the corner coefficients C_1, C_3, C_7, and C_9 are ignored.) For IPER = 1, periodic boundary conditions are used in the x direction (normal to the marching direction y). ICOR is the number of corrective clean-up iterations used to reduce round-off error accumulation; usually, ICOR = 0 is used, but in some cases of marginal stability, ICOR = 1 or 2 may be used.

The variable EMX is the output value of the maximum error in the solution, which occurs at the end of the march. A significant advantage of the basic marching method is that it will not lie to the user. The finite-difference stencil is satisfied virtually to single precision everywhere except at the end of the march. The solution obtained can be viewed as a virtually exact solution of a problem with a boundary condition perturbed by errors of order EMX or less.

The LU decomposition and backsolve of the influence coefficient matrix is done through LINPACK subroutines [6], which are selected by the option indicator NRCOND. For NRCOND = 1, LINPACK routines are used that give an estimate of the inverse of the condition number RCOND. The time penalty is small, and in my experience RCOND has been valuable as a debugging aid.

JMAR is an option indicator for the march direction, with ± 1 giving a marching the $+J$ or $-J$ direction, respectively. This is a significant option because the stability of the marching method is directional. For an expanding coordinate system (typical of boundary-layer calculations, for example), the stability is improved if the march proceeds from the coarse mesh to the fine mesh.

The next four arguments are primarily of use when GEM is driven by other codes, GEMPAT2 or GEMPAT4, which "stabilize" (or at least,"extend the range of") the solution by patching subregions together. Without stabilizing, JBOT = 1, JTOP = JL, NDBC = 0, and FDBC is ignored. In the stabilizing code, JBOT and JTOP define the extent of the subregion being solved, and NDBC = 1 indicates that the solution along the patching line has Dirichlet boundary conditions defined in the vector FDBC(*IL*).

IPVT and CI are work arrays, dimensioned IPV(*ILD*) and CI(*KLD*, *KLD*), where *KLD* ≥ IL - 2.

NC10 is another option indicator primarily used when patching subregions together. For NC10 = 0, the homogeneous problem $C10(I, J) = 0$ is solved, regardless of the values stored in $C10$.

Boundary conditions are also specified by the coefficients C_1–C_{10}, as indicated by the following stencils.

$$
\begin{array}{ccccc}
\begin{array}{c} C_5C_6 \\ C_8 \end{array} & \text{———} & \begin{array}{c} C_4C_5C_6 \\ C_8 \end{array} & \text{———} & \begin{array}{c} C_4C_5 \\ C_8 \end{array} \\
| & & & & | \\
\langle C_2 \rangle & & & & \langle C_2 \rangle \\
C_5C_6 & & & & C_4C_5 \quad F = C_{10} \\
(C_8) & & & & (C_8) \\
| & & & & | \\
\begin{array}{c} C_2 \\ C_5C_6 \end{array} & \text{———} & \begin{array}{c} C_2 \\ C_4C_5C_6 \end{array} & \text{———} & \begin{array}{c} C_2 \\ C_7C_8 \end{array}
\end{array}
\qquad (6.4.3)
$$

For example, this stencil indicates that in the lower left-hand corner, at $i = 1$ and $j = 1$, the boundary condition is

$$
C2(1,1) * F(1,2) + C5(1,1) * F(1,1) + C6(1,1) * F(2,1) = C10(1,1) \qquad (6.4.4)
$$

The general form of Eq. (6.4.4) allows for all linear combinations of boundary conditions such as Dirichlet, Neuman, mixed, ratio of derivatives ($aF_x + bF_y = c$), etc. However, the requirement for separability of boundary conditions in the marching y direction dictates that $C2$ cannot be used at the side boundaries for a march in increasing J (JMAR = +1), nor can $C8$ be used for JMAR = -1. In x, the periodic option indicator IPER = 1 overrides the matrix specification in Eq. (6.4.4). For the 9-point operator, the periodic tridiagonal solution is obtained by the method of Appendix B.

The generality in the boundary conditions does present a potential pitfall to the user. If the user specifies the problem in the coefficients $C1$–$C10$, including nonzero values for $C2$ and $C8$, then $C2$ will be ignored if JMAR = +1, and $C8$ will be ignored if JMAR = -1. Thus, the user will be solving two different problems depending on the march direction selected for the code. This potential pitfall is more serious for the patching solution. Along a patching line, we must have both $C2$ and $C8 = 0$. Again, this is caused by the requirement for the separability of the problem in the marching direction and along patching lines. Thus, if the user defines nonzero coefficients in all the arrays $C1$–$C10$, he or she will be solving a different problem depending on whether or not patching codes are used. If patching codes are used, the problem solved is that of $C2 = C8 = 0$ along the patching lines. Away from the patching lines, problems solved will depend on the marching direction JMAR in each of the subregions.

The safest approach to this pitfall is to solve only problems with $C2 = C8 = 0$. Indeed, we considered writing a version of the GEM code that solved only this class of problems, in order to avoid possible misuse. However, it was finally decided to include the nonzero values of $C2$ and $C8$, since this allows the specification of tangential derivatives at boundaries. This is a condition that does arise physically in problems in small disturbance transonic flow, in magnetohydrodynamics problems, and some other applications, and for any gradient boundary condition in non-orthogonal coordinates (see Section 1.3.7).

It should be noted that the significant storage problem of the ten arrays $C1$–$C10$ is not an aspect of the marching method, but simply follows from the problem description. Using arrays for all the coefficients in the stencil requires 10 arrays just to define the problem. The marching method itself (unstabilized) requires only an additional storage array for the influence coefficient matrix CI that is $IL \times IL$.

The code is written so that the smart user has the opportunity to save storage space by regenerating some or all of the two-dimensional coefficient arrays as external or statement FUNCTION's in FORTRAN. The user can remove an array such as *C*10, the right-hand side of the equation, from the common block GEMCOM and write a function subprogram with the same name and argument list; i.e., FUNCTION *C*10(*I*,*J*). (None of the arrays *C*1–*C*10 are passed to other subroutines in argument lists, and the unused portions of the arrays, e.g., *C*1 at $J = JL$, are not used for temporary storage.) To assist in such code modifications, and to facilitate the changing of dimension statements, the original code was written with the CDC UPDATE feature. Dimensions can be changed by referring only to a short section of code, which includes all DIMENSION specifications. In the later FORTRAN 77 version, this flexibility is provided via the INCLUDE statement.

6.5 Tests of the Basic GEM Code

One set of test problems used pseudo-random number generation for all coefficients, which was useful in debugging all the options. A second set used a simple Poisson equation modified by a cross-derivative term $VC * F_{xy}$ formulated with centered second-order differences.

The test problems in this chapter were run originally [1] on a now outdated "scientific" computer (CDC 6600; for Cray results, see Chapter 7), but the results are still of interest to users of more modern machines. The computer absolute speed is comparable to low-end workstations or high-end PC's. The single precision (SP) word length is slightly less than the double precision (DP) word length of a workstation or PC, so that the accuracy reported here is a conservative estimate for these machines. Since the computing environment now changes so quickly, *any* absolute timing tests would quickly become obsolete, so the more enduring information is the speed *relative* to SOR.

A sampling of the results is presented in Table 6.5.1. Note that the 81×81 mesh problem for the 5-point operator initiates in ~1 ms per cell, the equivalent of 64 point-SOR iterations, and solves repeat solutions in the equivalent of two point-SOR iterations. For the simple Poisson equation with $\Delta x/\Delta y = 10$, the maximum residual error is 3.9×10^{-6}.

The 9-point operator with nonperiodic boundary condition requires about 67% more initiation time and about 31% more repeat time. With periodic boundary conditions, the 9-point operator requires about 3.2 times as long for initiation and about 2.6 times as long for repeat solutions. The penalty for the repeat solution time could be reduced by saving the LU decomposition of the tridiagonal solver (Appendix B), but at the cost of an additional 2D work array.

These results were obtained on an early version of the basic GEM code and verify the operation counts of [1]. This version did not include the homogeneous override option needed for the stabilizing codes.

We have used an SOR iteration as a measure of work because of its popularity and ease of programming. There are, of course, more effective iterative methods. In retrospect, it would have been preferable to use the time to evaluate a residual for the basic "work unit", as done originally by Brandt [6]. For the 5-point operator, 1 SOR sweep is equivalent to about 1.5 work units based on residual evaluation.

	Problem Grid (operator)				
Timing quantity	31 X 31 (5 point)	51 X 51 (5 point)	81 X 81 (5 point)	81 X 81 (9 point)	Periodic 81 X 81 (9 point)
Init time	0.42	1.74	6.61	15.74	30.24
Init time/cell	0.47	0.70	1.03	2.46	4.73
Init time/SOR	29.7	42.8	64.3	107.1	205.7
Rep time	0.035	0.089	0.206	0.384	0.750
Rep time/cell	0.039	0.036	0.032	0.060	0.117
Rep time/SOR	2.48	2.19	2.00	2.61	5.10
Init time/rep time	12.0	19.6	32.1	41.0	40.3
% Error, θ_{rep}	-32	-22	-21	-28	-32
% Error, $\theta_{rep}/\theta_{rep}$	-22	-19	-15	-5	-3

Table 6.5.1 Timing Tests of the Early Version of the Basic Gem Code. Tests were performed on a CDC 6600 with level 2 optimization of the FORTRAN 4 code. Init. refers to initiation times, rep refers to repeat solution times, SOR refers to times for a single iteration of a point-SOR method including a convergence test but without boundary calculations, q refers to predictions based on theoretical operation counts. Total times are in seconds, times per cell are in millisecond, based on the minimum of three consecutive runs that included one initiation and one repeat.

6.6 The Stabilizing Codes GEMPAT2 and GEMPAT4

The method of stabilizing selected from several available alternatives (Chapter 3) is the multiple patching method. The problem size in J (i.e., maximum JL for given cell aspect ratio, etc.) is nominally doubled by breaking the solution into two subregions separated at $J = JPATCH$. With guessed Dirichlet boundary conditions at JPATCH, each subregion is solved directly, using basic GEM. This solution gives nonzero residuals along JPATCH. The new Dirichlet conditions along JPATCH are then solved directly so as to zero these residuals. The technique is a capacity matrix or influence coefficient matrix method, which is not essentially connected to marching methods. The patching matrix is established in an initiation procedure that requires IL - 2 homogeneous solutions with unit-perturbed Dirichlet conditions along JPATCH; hence, the requirement for the homogeneous override option NC10 in GEM.

This procedure for a two-patch solution is incorporated into the subroutine GEMPAT2, which then calls GEM. Although several of the options in GEM are not of interest except for use with GEMPAT2, it was decided to have only one version of GEM available. The possible confusion arising from the unused options seems outweighed by the advantage of having only one version of GEM to document and maintain. Similarly, GEMPAT4 implements the patching procedure for a four-patch solution, and it calls the only version of GEMPAT2.

The storage penalty for the stabilized versions becomes significant. The patching method for a two-patch solution requires two of the CI matrices (one for each subregion) and an additional storage penalty for the patching matrix, and so the storage penalty is $3 \times IL \times IL$, compared to $IL \times IL$ for the single-region solution by the basic GEM. For the four-patch solution, seven additional matrices are required. (If the problem was variables-separable, these storage penalties would be reduced to two and four matrices, respectively.)

6.7 Timing Tests of the Stabilized Codes

The timing tests for the stabilized versions of the GEM code are presented in Table 6.7.1. The zero-patch results in Table 6.7.1 differ from the results presented above in Table 6.5.1 for the earlier version of the code that did not have the option for homogeneous override. This option adds a multiplication and some other incidental subscripting to the operation count for the simple Poisson equation. This amounts to about a 25% penalty for the simple 5-point problem. For reasons of code management, the earlier (and faster) version of the basic GEM code was discontinued.

The timing tests used systematically refined square grids of 21 × 21, 41 × 41, 61 × 61, and 81 × 81. There is no essential difference between the 5-point operators for either aperiodic or periodic boundary conditions, but the condition of the 9-point operators significantly increases the time, and especially the 9-point periodic option. Asymptotically, the repeat time for large problems with a 5-point aperiodic operator is about 2 ½ SOR iterations. These data are for no corrective iterations (ICOR = 0). For a 101 × 101 grid (not shown in Table 6.7.1), the initiation time is about 111 SOR iterations, and the repeat time is 2.47 SOR iterations. The repeat time increases to 3.86 SOR iterations when one corrective iteration is used.

The additional time involved in the patching solutions is shown in Fig. 6.7.1, where the penalty ratio of the two-patch and four-patch solutions is compared to the zero-patch solutions. Depending on the problem solved (i.e., 5-point or 9-point, periodic or aperiodic boundary conditions), the penalty ratio is on the order of 2.2 for the two-patch solution with repeat solutions, and 4.4–6.8 for the four-patch solution with repeat solutions. The penalty ratio for initiation is approximately in the range 3–3.5 for the two-patch solution and 7.4–11.2 for the four-patch solution.

The operation count penalty for the two-patch and four-patch solutions was given in Section 3.4, but only for the constant-coefficient Poisson equation. The detailed operation count for the other operators has not been worked out, but the trends shown in Table 6.7.1 and Figure 6.7.1 do support the theoretical predictions. For the 5-point operator with nonperiodic boundary conditions on an 81 x 81 grid, the two patch code initializes in the equivalent of 334 SOR iterations, a factor of 3.7 over the single-region solution. This is somewhat *better* than the value of 4.3 predicted by the operation count. Repeat solutions are obtained in 5.7 SOR iterations, a factor of 2.3 over the single-region solution, in fair agreement with the value 2.1 predicted by the operation count. The four-patch solution initializes in 928 SOR iterations, a factor of 10.3 over the zero-patch solution, in fair agreement with the predicted factor of 11.5. The four-patch repeat solutions require 13.7 SOR iterations, a factor of 5.4 over the zero-patch solution, in fair agreement with the predicted factor of 4.4

The most convenient and enduring summary presentation of these timing results is given in terms of the work units measuring equivalent SOR iterations. These are presented in Table 6.7.2 for repeat times and initialization times. It is seen that even the four-patch solution is very competitive with other methods in terms of repeat times, giving repeat solutions for the 5-point operator in an 81 × 81 grid in the equivalent of 13.7 SOR iterations. However, the initialization time is deteriorating rapidly, requiring 928 SOR iterations to initialize. This, combined with the increasing storage penalties and the decreasing accuracy (see next section), indicates that higher patching beyond the four-patch is not a practical approach. (It is still faster than banded Gaussian elimination.)

The symmetry of marching methods is shown in Table 6.7.3, which compares solutions in a 21 × 61 grid with those in a 61 × 21 grid. The execution times for zero-patch repeat solutions are fairly symmetric, but are more asymmetric for the patched solutions and the initializations. Initialization of the 5-point operators with zero-patches in a 61 × 21 grid is more than twice that required for initialization in a 21 × 61 grid. The ratio for the 9-

point operator is a factor of 3. For the 5-point operator in a four-patch solution, the initialization time in the 61 × 21 grid is a factor of 5 greater than that in the 21 × 61 grid. For any particular physical problem, the errors are similarly asymmetric between these two problems.

Timing quantity	Problem Grid 21 X 21	41 X 41	61 X 61	81 X 81
A. Zero Patch, 5 point aperiodic operator				
Init time	0.237	1.63	5.25	12.1
Init time/cell	0.592	1.02	1.46	1.89
Init time/SOR	34.35	58.1	82.9	105
Rep time	0.025	0.082	0.168	0.285
Rep time/cell	0.021	0.017	0.016	0.015
Rep time/SOR	3.62	2.92	2.65	2.48
Init time/rep time	9.48	19.9	31.3	42.4
B. Zero Patch, 5 point periodic operator				
Init time	0.230	1.60	5.19	12.0
Init time/cell	0.575	1.00	1.44	1.88
Init time/SOR	33.3	56.8	82.0	107
Rep time	0.024	0.080	0.165	0.282
Rep time/cell	0.020	0.017	0.015	0.015
Rep time/SOR	3.78	2.84	2.61	2.50
Init time/rep time	9.58	20.0	31.5	42.6
C. Zero Patch, 9 point aperiodic operator				
Init time	0.321	2.39	7.87	18.4
Init time/cell	0.803	1.49	2.19	2.88
Init time/SOR	35.7	64.7	93.8	123
Rep time	0.035	0.120	0.256	0.446
Rep time/cell	0.029	0.025	0.024	0.023
Rep time/SOR	3.89	3.25	3.05	2.97
Init time/rep time	9.17	19.9	30.8	41.3
D. Zero Patch, point periodic operator				
Init time	0.574	4.29	14.2	33.2
Init time/cell	1.44	2.68	3.94	5.18
Init time/SOR	63.2	115	169	221
Rep time	0.059	0.219	0.469	0.809
Rep time/cell	0.049	0.046	0.043	0.042
Rep time/SOR	6.48	5.90	5.58	5.39
Init time/rep time	9.75	19.6	30.2	41.0

Table 6.7.1 Timing Tests of the Stabilized GEM Code. Tests were performed on a CDC 6600 with level 2 optimization of the FORTRAN 4 code. Init. refers to initiation times, rep refers to repeat solution times, SOR refers to times for a single iteration of a point-SOR method including a convergence test but without boundary calculations, q refers to predictions based on theoretical operation counts. Total times are in seconds, times per cell are in millisecond, based on the minimum of three consecutive runs that included one initiation and one repeat.

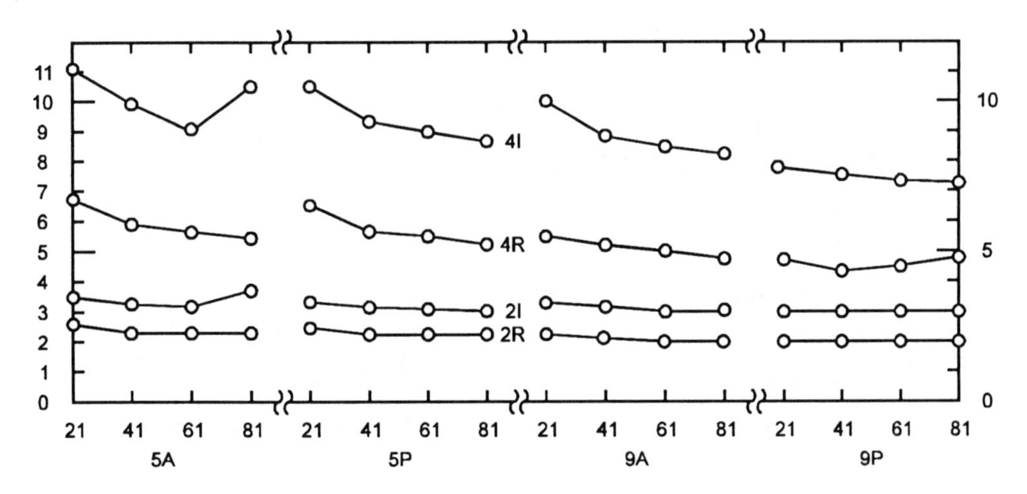

Figure 6.7.1. Penalty ratio for patching solutions compared to the zero-patch code. Numerals 2 and 4 refer to two-patch and four-patch; I and R refer to initialization and repeat solutions; 5 and 9 refer to 5-point and 9-point operators; A and P refer to aperiodic and periodic boundary conditions in I.

Timing quantity		Problem Grid (operator)			
		21 X 21	41 X 41	61 X 61	81 X 81
		A. Repeat times (rep time/ SOR)			
Zero Patch,	5-A	3.62	2.92	2.65	2.53
	9-A	3.89	3.25	3.05	2.97
	9-P	6.48	5.90	5.58	5.39
Two Patch,	5-A	9.70	6.85	6.25	5.73
	9-A	9.00	7.28	6.71	6.49
	9-P	13.8	12.1	11.5	11.2
Four Patch,	5-A	24.9	17.3	15.1	13.7
	9-A	22.7	17.2	15.2	14.7
	9-P	30.4	26.0	24.7	23.8
		B. Initialization times (init time/ SOR)			
Zero Patch,	5-A	34	58	83	90
	9-A	36	65	94	123
	9-P	63	116	169	221
Two Patch,	5-A	126	194	267	334
	9-A	121	208	294	382
	9-P	193	348	507	663
Four Patch,	5-A	391	567	759	928
	9-A	363	575	774	1008
	9-P	494	876	1252	1640

Table 6.7.2 Timing Tests of the Stabilized GEM Codes Measured in Equivalent SOR Iterations, including Convergence Tests. 5 and 9 refer to 5-point and 9-point operators; A and P refer to aperiodic and periodic operators.

	Problem Grid			
	5-A		9-P	
Timing Quantity	21 X 61	61 X 21	21 X 61	61 X 21
Zero-patch reptime/SOR	30	65	60	180
Two-patch	96	367	178	569
Four-patch	256	1313	428	1593
Zero-patch reptime/SOR	3.08	3.30	6.19	6.23
Two-patch	6.75	9.51	12.6	13.4
Four-patch	15.7	27.4	26.0	31.4

Table 6.7.3 Asymmetry of the Timing Tests; See Table 6.7.2 for explanation.

6.8 Representative Accuracy Testing

The accuracy testing of the GEM codes is difficult to generalize because of the generality of the problems being solved and the sensitivity of the method to these problems. Since the code is capable of handling coefficient matrices that are generated by random number generators, it is obviously impossible to assign a specific accuracy to the code. Representative accuracies are obtained by solving the simple Poisson equation in Cartesian coordinates with or without the addition of a small coefficient (0.1) for a cross-derivative term. Simple Dirichlet and periodic boundary conditions were used for these tests. The problem size and the cell aspect ratio $\beta = \Delta x/\Delta y$ were systematically varied.

As noted in Chapter 1, Section 1.2.5, the cell aspect ratio β is the dominant parameter in these tests. For example, for a 15×15 grid problem with $\beta = 1$, the GEM codes give a maximum error of 6.2×10^{-5} with no corrective iterations being used. For a 101×101 problem, the use of $\beta = 1$ will result in overflows in the computer. However, an aspect ratio $\beta = 10$ gives a maximum error of 2.4×10^{-3}. An aspect ratio of 15 in this 101×101 grid problem gives an error of 7.5×10^{-6}.

For these accuracy tests, the grid was systematically varied over $IL, JL = 15, 29, 57$. For the patching tests, the patching lines are approximately equally distributed. That is, for the problem with $JL = 15$, the patching lines were at $J = 4, 8, 11$. For $JL = 29$, the patching lines were at $J = 8, 15, 22$. For $JL = 57$, the patching lines were at $J = 15, 29, 43$.

The effect of number of patches is shown in Table 6.8.1. The use of the patching code increases the accuracy in all cases, although not as much as anticipated. Nominally, the patching solution should double the size of the problem for a given accuracy. In fact, this is true only in the limits of small IL, i.e., a one-dimensional problem in J. For the two-dimensional problems, the error in the solution of the patching matrices interacts with the marching error in each subregion. If the patching procedure were actually error-free, the error for a zero-patch solution at $JL = 15$ would be of the same order as the errors for a two-patch solution at $JL = 29$ and for a four-patch solution at $JL = 57$. In fact, as seen from inspection of Table 6.8.1, the error deteriorates as the mesh is doubled and a patching solution is used. For $IL = 15$, the solutions for the sequence (zero-patch, $JL = 15$), (two-patch, $JL = 29$), (four-patch, $JL = 57$) give maximum errors of 6.2×10^{-5}, 4.4×10^{-4}, and 1.5×10^{-3}, respectively, instead of being approximately constant. The deterioration is even more rapid for the 9-point operator with periodic boundary conditions.

	ICOR						
	0	1	0	1	0	1	Patch
A. Five-point aperiodic operator with $\beta = 1$							
	-	-	-	-	-	-	0
57	-	-	-	-	-	-	2
	1.5E-03	1.9E-03	5.8E-03	7.4E-03	1.6E-02	1.6E-02	4
	-	-	-	-	-	-	0
29	4.4E-04	1.2E-03	2.1E-03	4.8E-04	3.5E-03	6.4E-04	2
	9.3E-09	8.9E-09	2.4E-08	3.0E-08	3.7E-08	3.7E-08	4
	6.2E-05	3.7E-04	2.0E-04	1.3E-04	1.5E-04	8.5E-05	0
JL = 15	8.9E-10	5.0E-10	2.3E-09	1.3E-09	4.0E-09	1.8E-09	2
	3.0E-11	1.6E-11	3.0E-11	7.1E-11	5.5E-11	7.2E-11	4
	IL = 15		29		57		
B. Nine-point aperiodic operator with $\beta = 1.5$							
	-	-	-	-	-	-	0
57	-	-	-	-	-	-	2
	5.8E-06	6.9E-06	5.8E-03	7.4E-03	4.1E-05	3.7E-05	4
	-	-	-	-	-	-	0
29	2.8E-06	2.1E-07	8.0E-06	1.9E-05	1.1E-05	3.0E-06	2
	7.4E-10	3.7E-10	1.9E-09	2.1E-09	1.9E-09	2.1E-09	4
	1.9E-07	8.2E-08	3.2E-07	1.1E-07	5.7E-07	2.4E-07	0
JL = 15	2.0E-10	9.2E-11	3.6E-10	1.8E-10	3.6E-10	1.9E-10	2
	6.9E-12	7.3E-12	1.1E-11	1.8E-11	2.2E-11	2.1E-11	4
	IL = 15		29		57		
C. Nine-point aperiodic operator with $\beta = 2.0$							
	-	-	-	-	-	-	0
57	-	-	-	-	-	-	2
	3.1E-03	5.3E-05	9.2E-03	9.4E-05	1.4E-02	9.2E-05	4
	-	-	-	-	-	-	0
29	1.5E-03	3.0E-07	2.9E-03	1.3E-06	3.9E-03	1.9E-06	2
	1.1E-08	7.3E-10	1.0E-08	5.7E-09	6.0E-09	4.5E-09	4
	5.7E-06	9.7E-09	1.4E-04	7.1E-09	2.2E-04	8.0E-09	0
JL = 15	3.5E-09	2.2E-09	1.3E-09	2.5E-10	1.2E-09	1.6E-10	2
	2.9E-11	2.6E-11	1.1E-11	9.7E-12	1.9E-11	1.1E-11	4
	IL = 15		29		57		

Table 6.8.1 Accuracy Tests on a Simple Poisson Equation with an Additional Term 0.1 $\delta^2 F/\delta x \delta y$ for the 9-Point Operators; Table Shows Maximum Residual Error.

The effect of number of corrective iterations ICOR on the accuracy of repeat solutions is somewhat unpredictable. The effect on the timing tests is very simple, with a 50–60% penalty accruing for ICOR = 1, independent of the problem size and of the number of patching regions. For the 5-point and 9-point aperiodic operators, the use of ICOR = 1 usually results in a slight improvement in maximum error. However, in some cases the accuracy is slightly deteriorated. (In no case will a number of corrective iterations actually cure a very inaccurate problem.) However, for the 9-point periodic operator, the use of ICOR = 1 can dramatically improve the accuracy. For example, in the 29 × 29 grid problem with a two-patch solution, the use of ICOR = 1 results in the maximum error decreasing from 2.9×10^{-3} to 1.3×10^{-6}.

Generally, use of one corrective iteration seems advisable only for the 9-point periodic operator. (It should be noted that in many applications in fluid dynamics and heat transfer, the GEM codes will be used repetitively and an excellent initial condition might be available from a previous iterative solution of a nonlinear or time-dependent problem. In such a case, the good initial conditions have the same effect as using ICOR > 0 in the present tests.)

Unlike the single-region solution, in which the error is virtually confined to the boundary at the end of the march, the patched solutions also have errors (nonzero residuals) along and adjacent to the patching lines. However, the patching matrix is usually well-conditioned and this error is acceptable in the problems tested to date.

6.9 Conclusions

Timing and accuracy testing of the GEM codes, a particular realization of marching methods for elliptic equations, has been presented. The codes are written for a "smart user" and treat 5-point and 9-point nonseparable operators with a variety of boundary conditions. The codes also include an option for the user to select the marching direction in all of the subregions, an option that is important for expanding mesh solutions such as those typical of boundary layer calculations. The codes are written so that any of the matrices that describe the problem may be easily replaced with statement functions or external functions, so as to save computer memory. There are pitfalls in the use of the codes, particularly the stabilized codes.

The basic (nonstabilized) GEM code depends on the problem parameters, primarily a large cell aspect ratio $\Delta x/\Delta y$, to control the instability incurred in marching the elliptic equations. For a 5-point operator with nonperiodic boundary conditions, repeat solutions for large problems are indeed obtained in a *time equivalent of two point-SOR iterations.* Treatment of 9-point operators and periodic boundary conditions increase the execution time, but repeat solution times are still optimal, i.e., merely proportional to the number of unknowns. With a large cell aspect ratio, problems as large as 101×101 have been solved accurately with this code.

Stabilizing codes that patch together subregions solved by the basic GEM code nominally double and quadruple the problem size in the marching direction. However, some degradation in the accuracy is noted, apparently due to interaction of the rounding errors from the marching method and from the solution of the patching matrix. The timing penalty for the mesh-doubling and mesh-quadrupling codes is reasonable for repeat solutions, but becomes a serious disadvantage for initialization as the number of patching regions increases. The rapidly deteriorating initialization time, decreasing accuracy (compared to the nominal), and rapidly increasing storage penalty for the patching matrices indicate that the patching algorithms become impractical beyond the four-patch solution. The maximum problem size remains strongly dependent on a favorable cell aspect ratio.

A single corrective iteration, which increases the repeat solution time by 50–60%, markedly improves the accuracy of the 9-point periodic operator, and is recommended as standard use for this problem. For other problems, the corrective iteration has a somewhat unpredictable effect, usually giving a small increase in accuracy but sometimes giving a decrease.

For the basic GEM code, the error in the residual is isolated at the boundary at the end of the march. This means that the code gives an essentially exact solution of a problem with perturbed boundary conditions. When the patching solutions are used, the codes also produce nonzero residuals at interior points at and adjacent to the patching lines.

Carefully performed operation counts of the algorithm have been proven to be a dependable and fairly accurate indicator of the relative computational speed of the codes.

References for Chapter 6

1. P. J. Roache, Performance of the GEM Codes on Non-Separable 5- and 9-Point Operators, *Numerical Heat Transfer*, Vol. 4, No. 4, 1981, pp. 395-408.
2. P. J. Roache, Marching Methods For Elliptic Problems: Part 1, *Numerical Heat Transfer*, Vol. 1, 1978, pp. 1-25; Part 2, Vol. 1, 1978, pp. 163-181; Part 3, Vol. 1, 1978, pp. 183-20.
3. P. N. Swarztrauber and R. A. Sweet, Efficient FORTRAN Subprograms for The Solution of Separable Elliptic Partial Differential Equations, *Transactions in Mathematical Software*, Vol. 5, 1979, pp. 352-364. The software is available from National Center for Atmospheric Research, Boulder, CO. A commercial software package of vectorized codes is available from Green Mountain Software, 1951 Alpine Avenue, Boulder, CO 80304.
4. P. J. Roache, A Sixth-Order Accurate Direct Solver for the Poisson and Helmholtz Equations, *AIAA Journal*, Vol. 17, 1979, pp. 524-526.
5. B. Parlett, Progress in Numerical Analysis, *SIAM Review*, Vol. 20, 1978, pp. 443-458.
6. J. J. Dongarra, C. B. Moler, J. R. Bunch, and G. W. Stewart, *LINPACK User's Guide*, Society for Industrial and Applied Mathematics, Philadelphia, 1979.

Chapter 7

VECTORIZATION AND PARALLELIZATION

7.1 Introduction

In this chapter, we address the questions of how well the marching methods vectorize and parallelize [1]. The answers will depend strongly on what type of problem is being solved, and on the particular parallel computer being considered. The marching methods do not vectorize very well for the 9-point operator, but vectorize very well indeed for the 5-point operator. For repeat solutions of the 5-point operator, the combination of an optimal scalar operation count for the marching methods (see Chapter 1, Section 1.2.4, Chapter 6, and [2,3]) and near super-vectorizing produces solution times that are very fast indeed. As discussed in Section 6.5, since modern computing environments change so quickly, *any* absolute timing tests would quickly become obsolete. Also, research on the scalability of marching methods remains to be done, but relative results from [1] presented herein still verify the levels of vectorization achieved.

7.2 Vectorizing the Tridiagonal Algorithm and the 9-Point March

The vectorizing for the 9-point algorithm is limited by the vectorizing of the tridiagonal algorithm required for every line of the march, both in the initialization stage to establish the influence coefficient matrix CI, and for the two marches of the repeat solutions (Chapter 1, Section 1.3.6).

A great deal of work has gone into attempts to vectorize the tridiagonal algorithm [4,5]. For use in ADI and LSOR methods, it is easy and profitable to run parallel computations for simultaneous equations. For example, in an IL $\times$ JL grid, with L = IL - 2 and M = JL - 2, all L equations in the J direction, each JL long, can be solved in parallel by lagging the dependence on (I $\pm$ 1) neighbors or by using alternate line ordering. However, the applications to marching methods require each single equation, L long, to be solved sequentially. The Gaussian elimination code for this problem (i.e., the Thomas algorithm) has some vectorizable parts, but some that are inherently not vectorizable. Some increase in speed is obtained using a short divide [4], with only 9 significant decimal figures of accuracy on the Cray-1, but this would not be satisfactory for the precision sensitive marching methods. The cyclic reduction method vectorizes, but has a much higher scalar operation count, which also hurts the error accumulation, important in the application to marching methods. A polyalgorithmic approach is apparently optimum, depending on the size of the problem [4]. Some vectorization occurs, but in the range of problem sizes generally of interest for marching methods (say IL = 50 to 150), the best performance obtained by Jordan [4] gave a factor of 2 or perhaps 2½ increase over scalar Gaussian elimination. Even this modest vectorization is obtained at the cost of loss of precision (not important for iterative methods such as ADI and LSOR), with the removal of pivoting (not important for marching methods on the simple Poisson equation, but possibly important for more general problems), and with the effort of assembly language programming.

As Jordan concludes [4], "pipeline computers appear to offer little advantage over scalar algorithms for this problem." The vectorizing would be somewhat improved when a higher order equation, such as the fourth-order biharmonic equation or the coupled second-order equations were being solved (Chapter 2, Section 2.4) since the vectorizing performance of the Cray-1 and other vectorizing computers improves as we increase the bandwidth of the matrix [4]. But generally, we can conclude that the tridiagonal algorithm, and therefore marching methods for the 9-point operator, do not vectorize well.

7.3 Vectorizing the 5-Point March

The situation is radically different for the 5-point operator, which does not require a tridiagonal solution for the march. The marching part of the algorithm vectorizes with a vector length = L = IL-2. To achieve significant vectorization, this vector length L should be at least as large as the vector half-performance length of the computer, $n_{½}$. Hockney [5] has defined $n_{½}$ as that vector length which achieves half of the asymptotic performance of the parallel computer, i.e., half of that performance that accrues to the largest possible length vector operation. The value of $n_{½}$ for any particular computer varies somewhat with the particular floating point operations being performed, but for a Cray-1 computer it is typically in the range of 10 to 20. It is thus easy to achieve significant vectorization of the 5-point marching algorithm with a reasonably sized problem on the Cray-1. This vectorization is not limited to the pipeline architecture of the Cray-1, but would also accrue with a processor array machine with the same natural hardware parallelism and same value of $n_{½}$. On the other hand, the old Cyber 205 had $n_{½}$ of about 100, and the ICL DAP had $n_{½}$ of 2048, making the marching methods useless for its parallel capability.

The Gaussian elimination routines (LU decomposition and backsolve) that are part of the marching algorithm also vectorize significantly, though not as completely as the 5-point march.

7.4 Timing and Accuracy for the Vectorized Marches

Although the emphasis in Chapter 6 was on the variables nonseparable problem, we decided to prove the vectorizing capability of the marching methods [1] on a stripped-down version of the GEM code restricted to the constant coefficient 5-point stencil obtained by central differencing of the Poisson equation in cartesian coordinates, with simple Dirichlet boundary conditions (specified function values). Only the basic marching method was used, i.e., no stabilization, so that accuracy depends on a large (though still practical) cell aspect ratio = 10.

Timing and accuracy results are given in Table 7.4.1 for a fast scalar machine, the Cyber 760, and in Table 7.4.2 for the pipeline vector Cray-1S. Both machines used were at Boeing Computer Services. (Although the Cyber 760 was a scalar machine in that it did not have the vector registers like the Cray-1, it actually utilized functional parallelism in its arithmetic; see e.g. [5].) For the 9-point operator, the results (not shown) indicated about a factor of 3 between these two machines. A factor of 1.75 [5] or 2 [4] is typical for entirely scalar operations, indicating less than a factor of 2 improvement due to vectorization, as expected from the discussion above on the tridiagonal problem.

The asymmetry index in the Tables shows the influence of the directionality of the marching method, and is generally consistent with the scalar operation counts given in Chapter 1 and in [2].

The difference in the maximum errors on the two computers was not due to the differences in word length. The Cyber 760 used a 60-bit word and the Cray-1 used a 64-bit word, but the additional 4 bits on the Cray-1 were not used in the mantissa of the floating point representation (unfortunately, since this would increase the precision) but in a larger characteristic. Rather, the difference was due to a difference in the floating point division arithmetic on the two machines. (The Cray-1 actually solved floating point division by a hardware Newton-Raphson iteration, which is vectorizable. The division operations do not occur in the march for the simple 5-point stencil, but in the Gaussian elimination routines.)

Note that the repeat times per cell actually decrease as size increases, implying a better-than-optimal operation count because of overhead operations. (See also the 6th-order special cylindrical Poisson solver by the present author in [6].)

The optimum value of L for the Cray-1 would be L = 64 (its pipeline vector length) or integer multiples thereof, giving IL = 66 and 130. However, we chose instead IL=65 and 129 for easier comparison with methods using FFT and/or cyclic reduction. The performance for IL = 66 is only slightly better (less than 1 percent) than IL=65 on a per cell basis. However, the change from IL = 66 to IL = 67 increases the repeat solution time per cell by 13 percent, and a comparable increase for initialization. This step increase is due to the reloading of the vector registers, according to Hockney and Jesshope [5, p.91]. In a simple DO loop for the element-by-element multiplication of two vectors, their results show steps of about 0.5 microseconds (about 14 percent) at increments of 64. They also attribute a 1 microsecond start-up time to initiate the loop to compiler overhead which could be avoided in careful assembler code.

5-Point constant coefficient Poisson operator, $\Delta x/\Delta y = 10$				
mesh size	65×65	65×129	129×65	129×129
max error	$1.8e^{-10}$	$2.2e^{-05}$	$2.0e^{-10}$	$2.9e^{-05}$
init time with rcond	0.669	1.280	2.682	5.171
rcond	$7.6e^{-05}$	$4.5e^{-10}$	$7.4e^{-05}$	$4.3e^{-10}$
init time (w/o rcond)	0.660	1.275	2.650	5.085
init time/cell	0.161	0.156	0.323	0.310
rep time	0.0174	0.0312	0.0361	0.0639
rep time/cell	0.00425	0.00381	0.00441	0.00390
init time/rep time	37.9	40.9	73.4	79.6
asymmetry index = time (129×65)/time(65×129) init = 2.650/1.275= 2.088 rep= 0.0361/0.0312= 1.16				

Table 7.4.1 CYBER 760

5-point constant coefficient Poisson operator, $\Delta x/\Delta y = 10$				
mesh size	65×65	65×129	129×65	129×129
max error	$2.0e^{-10}$	$3.4e^{-05}$	$1.8e^{-10}$	$3.1e^{-05}$
init time with rcond	0.052	0.099	0.186	0.355
rcond	$7.6e^{-05}$	$4.5e^{-10}$	$7.4e^{-05}$	$4.3e^{-10}$
init time (w/o rcond)	0.050	0.097	0.182	0.351
init time/cell	0.012	0.012	0.022	0.021
rep time	0.00122	0.00222	0.00233	0.00414
rep time/cell	0.000300	0.000271	0.000284	0.000253
init time/rep time	41.1	43.8	78.2	84.8
MFLOP/S, init	28.5	27.8	35.6	33.1
MFLOP/S, repeat	45.5	46.8	55.1	54.5
asymmetry index = time (129×65)/time(65×129) init = 0.182/0.097= 1.88 rep= 0.00233/0.00222= 1.05				

Table 7.4.2 CRAY-1S

7.5 Efficiencies

There are two efficiencies involved in parallel computations, the scalar efficiency of the algorithm, which is indicated approximately by operation counts such as those given previously (in Chapter 1, Section 1.2.4) and the efficiency of utilization of the parallelism of the computer, indicated indirectly by calculating its operating speed in MFLOP/S, or millions of floating point operations per second. These two efficiencies are often opposed. For example, a very inefficient algorithm such as Jacobi (Richardson) iteration [7] makes very efficient use of parallelism. This trade-off between algorithmic inefficiency and machine efficiency is a major theme of computational research. However, in the present case, these is no trade-off, and the result is a very efficient code by any criterion.

Jordan [4] (see also [5]) distinguishes three rough levels of vectorization on the Cray-1 computer. "Scalar" is a performance level that does not use vector operations; he estimates that codes at this "Scalar" level will perform about two times faster than the old CDC 7600. (Hockney and Jesshope [5] also estimate about a factor of 1.75.) "Vector" is a performance level that uses vector operations in which memory references are the limiting factor. (The low memory bandwidth of 80 Mwords/second is the principal bottleneck of the Cray-1 [5].) Jordan [4] estimates that an approximate 4 to 1 performance ratio between the Cray-1 and the old CDC 7600, (or about 2 times the "Scalar" level) "seems reasonable" for this "Vector" level of vectorization. "Super Vector" is a performance level that uses vector operations in which performance is limited by the availability of functional units and/or vector registers. Super Vector performance is attainable in algorithms in which a considerable amount of arithmetic is done on a few operands, so that the vector registers can serve as the primary memory. He estimates Super Vector performance level to be about three times faster than that of the "Vector" level (or about six times the "Scalar" level). Hockney and Jesshope [5] also use Jordan's categorization of performance. In their examples, the actual Cray-1 performance varied from 2.5 to 153 MFLOP/S; the theoretical limit of performance is about 151 MFLOP/S.

The MFLOP/S for the present timing tests [1] were calculated from the measured run times and the number of scalar operations, determined from the operation counts in Chapter 1, Section 1.2.4. For initialization, for the 5-point operator from Eqs. (1.2.25 - 1.2.34), the L marches to establish CI require $2ML^2$ multiplies and $3ML^2$ adds. In this special code, we did not program the variable loop index, which avoids up to 1/3 of the null calculations in these marches, because the additional complexity would reduce the vectorizing. Hence, we use the factor $\lambda = 1$ in Eq. (1.2.26). The LU decomposition requires $L^3/3$ multiplies and adds, plus lower order terms. For L = M, the approximate total number of floating point operations is thus 5 $2/3M^3$ of which $5M^3$ are for the marches to establish CI, and $2/3M^3$ are for the LU decomposition. For repeat solutions, the two marches (assuming no corrective iterations) require 4LM multiplies and BLM adds, and the LU backsolve requires L^2 multiplies and adds. For L = M, this gives a total of 14M*2 floating point operations, $12M^2$ for the two marches and $2M^2$ for the LU backsolve. Initialization in a 65 × 65 grid thus involves 5 2/3 (63^3) = 1.42E+0.6 floating point operations. The repeat solutions in a 65 × 65 grid then involve $14(63^2)$ = 55,566 floating point operations. Divided by the run time, these give the total output of the machine in MFLOP/S. Actual entries in Table 7.4.2 were based on operation counts without the assumption of L = M, and included lower order terms.

As shown in Table 7.4.2, for this stripped-down GEM code, the Cray-1S was running at 28 to 36 MFLOP/S during the initialization, and at 46 to 55 MFLOP/S for repeat solutions.

For further comparison, we note the result quoted in [5], p. 91, for vectorized Fortran code on a Cray-1S for the simple problem of element-by-element multiplication of two one-dimensional vectors, which gave an asymptotic rate of 22 MFLOP/S.

This near "Super Vector" level of performance for the stripped-down GEM code was obtained for a predominately Fortran code, with no overt effort at vectorization. The LU decomposition and backsolve subprograms used by the code had been extensively optimized and vectorized by Boeing Computer Services personnel, including considerable assembly language programming in CAL (Cray Assembly Language). But these portions only represent approximately 2/17 or 12 percent of the scalar operation count for initialization, and 1/7 or 14 percent for repeat runs, for L = M. The Fortran code for the marching method *per se* vectorized automatically using CFT, the Cray Fortran compiler.

It is noteworthy that CFT recognized the distinction between rows of a doubly-subscripted array. In the 5-point march equation, Eq. (1.2.17), F(I, J+1) appears on the left of the replacement statement. On the right, the second subscripts are only J and J-1. CFT vectorized this statement as though F(I, J+1) were a one-dimensional vector array in I, independent of F(I,J) and F(I,F-1) on the right hand side. (In fact, efforts to help the vectorizing by using separate one dimensional arrays for the left hand side of the march equation only slowed down the code because of the additional replacement statement needed.) Because the inner loop of the repeat solution vectorized so easily (and the Cray-1 can vectorize only innermost loops) it is doubtful that assembly language coding would have improved the performance very significantly; our guess is 20 percent maximum.

The initialization was performed using the general matrix routines from LINPACK [8]. As pointed out in Chapter 1, Section 1.2.4, for the constant coefficient Poisson problem considered here, about half of the scalar operation count for the LU decomposition, representing about 1/17 or 6 percent of the total initialization for L = M, can be saved by using a symmetric matrix solver without pivoting. It is known from experience that the pivoting is not necessary for this well-behaved equation (although for more general equations, likely for those whose matrix is not diagonally dominant, the pivoting would be important). When the corresponding symmetric matrix solver of LINPACK was used, a slight *increase* in the execution time occurred, instead of the expected decrease. The reason is that, at the particular computer facility used (Boeing Computer Services), a considerable effort had been expended in optimizing and vectorizing the general matrix routine, including assembly language programming, but this effort had not been expended on the symmetric matrix routine. This effort more than made up for the factor of 2 difference in the scalar operation count. If the same effort were expended on the symmetric matrix routines, presumably the initiation times given in Table 7.4.2 would be decreased by about 5 - 6 percent.

The combination of optimal scalar operation count for repeat solutions (i.e., operation count merely proportional to the number of unknowns) and near "Super Vector" level of machine utilization gives a code that is very fast indeed. Repeat solutions in a 65 $\times$ 65 grid, with almost 4,000 coupled interior equations, are obtained in 1.2 milliseconds each. For the 129 $\times$ 129 grid, with over 16,000 coupled interior equations, repeat solutions are obtained in 4.1 milliseconds. These were apparently the fastest solutions obtained for this type of problem at the time of publication of [1].

7.6 Multiprocessor Architectures

As computer architectures become more exotic, the performance of the ensemble of computer/language/algorithm/problem becomes more difficult to predict and more sensitive to changes in any parameters. We have not yet done any testing on multiprocessor computers, nor any detailed operation counts, but the following general observations are probably valid. (See also the discussion in the introduction to Chapter 3.) We assume an ideal multiprocessor computer architecture in which each processor can be viewed as a full computer (i.e., MIMD architecture) each of which is powerful enough to accommodate the marching methods.

The simple 2-D march of the 5-point equation (Section 1.2.2) could be set up in a natural parallelism with a separate processor for each i-th column. However, the need for data transfer from both the neighboring processors at i $\pm$ 1, and the small number of arithmetic operations relative to the number of data transfers, indicate that such an arrangement would not achieve high efficiency for the stripped-down cartesian Poisson problem, but that the general coefficient 5-point operator, with its greater arithmetic for the same amount of data transfer, would be better. The 2-D march of the 9-point operator would be much worse, since that problem does not have a natural parallelism. The same observations apply to the Simple 3-D marching method (Section 5.2).

The best candidates for multiprocessor parallelism would be stabilizing algorithms with many subregions, i.e., Domain Decomposition (Sections 3.3-3.6, 3.8, 5.12), and the 3-D EVP-FFT method (Section 5.8). The march in each subregion of the stabilizing algorithm could be performed on a separate processor, with relatively low data transfer requirements. For the 3-D EVP-FFT method each Fourier component could be marched out on a separate processor. This 3-D problem would be efficiently implemented on a multiprocessor computer.

As noted earlier (Chapter 3, Section 3.9) the basic and stabilized Marching Methods are especially suitable for serial high-end PC's and workstations. This means that they are also suitable for *virtual parallel networks*, in which heterogeneous nodes on a computer network are utilized as available via software such as PVM [9] and P4 [10]. Also, preliminary experiments by J. Morris and D. Keyes [11] indicate that simple marching methods can be quite competitive in Domain Decomposition as preconditioners for a Krylov-Schwarz algorithm.

References for Chapter 7

1. P. J. Roache, Additional Performance Aspects of Marching Methods for Elliptic Equations, *Numerical Heat Transfer*, Vol. 8, 1985, pp. 519-535.
2. P. J. Roache, Marching Methods For Elliptic Problems: Part 1, *Numerical Heat Transfer*, Vol. 1, 1978, pp. 1-25, 1978; Part 2, *Numerical Heat Transfer*, Vol. 11, 1978, pp. 163-18; Part 3, *Numerical Heat Transfer*, Vol. 1, 1978, pp. 183-201.
3. P. J. Roache, Performance of the GEM Codes on Nonseparable 5- and 9-Point Operators, *Numerical Heat Transfer*, Vol. 4, 1981, pp. 395-408.
4. T. L. Jordan, A Guide to Parallel Computation and Some Cray-1 Experiences, *Parallel Computations*, G. Rodriguez (ed.), Academic Press, New York, 1982, pp. 1-50.
5. R. W. Hockney and C. R. Jesshope, *Parallel Computers*, Adam Hilger Ltd., Bristol, England, 1981.
6. P. J. Roache, A Sixth-Order Accurate Direct Solver for the Poisson and Helmholtz Equations, *AIAA Journal,* Vol. 17, 1979, pp. 524-526.
7. P. J. Roache, *Computational Fluid Dynamics*, Hermosa Publishers, Albuquerque, NM, rev. printing, 1976.
8. J. J. Dongarra, C. B. Moler, J. R. Bunch, and G. W. Stewart, *LINPACK User's Guide*, Society for Industrial and Applied Mathematics, Philadelphia, 1979.
9. A. Geist, A. Beguelin, J. Dongarra, W. Jiang, R. Manchek, and V. Sunderam, *PVM 3 User's Guide and Reference Manual*, Oak Ridge National Laboratory, TN, ORNL/TM-12187, May 1993.
10. R. Butler and E. Lusk, *User's Guide to the p4 Parallel Programming System*, Argonne National Laboratory, IL, ANL-92/17, Oct. 1992.
11. J. D. Morris and D. E. Keyes, Marching Preconditioners for Krylov-Schwarz Methods in the Solution of Transport Problems, Feb. 1994.

Chapter 8

SEMIDIRECT METHODS FOR NONLINEAR EQUATIONS OF FLUID DYNAMICS

8.1 Introduction: Time-Dependent Calculations versus Semidirect Methods

In this chapter, we consider iterative application of Marching Methods to nonlinear partial differential equations in "Semidirect" methods. These methods are good candidates for parallel solutions of nonlinear problems via Domain Decomposition.

The example problems herein are taken primarily from fluid dynamics. Many of the problems of interest in fluid dynamics are essentially transient. For example, weather predictions, oscillating rotors, flow around heart values, etc., are problems in which a "steady-state" solution is of no relevance. For such problems, transient numerical methods are obviously required, and there are a variety of methods to choose from (e.g., see Roache [1]).

There are also many problems of practical interest where the steady-state or quasi-steady-state solutions are of interest. Historically, the most popular and dependable approach to obtaining a steady-state solution has been to mimic nature by solving a transient or pseudo-transient problem, obtaining the steady solution asymptotically in time. However, this approach can be very costly, especially for fine-grid resolution. For hundreds of computational cells in each spatial direction in finite difference solutions, it may well take hundreds or even a thousand time-steps to reach a steady state.

The Semidirect methods described in this chapter can obtain steady-state solutions in roughly of 10 iterations. Each of these iterations requires a computer time comparable to a time-step in the *more efficient* of the time-dependent methods (those that use an efficient solver for an elliptic equation within the time step). Compared to the *less* efficient of the time-dependent methods (those that involve a nested inefficient iterative solver for an elliptic equation within each time step), the Semidirect methods can actually attain a complete steady-state *solution* in computer time comparable to that for a *single time-step*. This represents a gain of 2 or 3 orders of magnitude in computing speed.

One incontrovertible advantage remains for the time-dependent methods — they do not require the assumption that a steady-state solution exists and is stable. Even this advantage is somewhat ameliorated because the time-dependent or time-like methods that are best suited to steady-state calculations introduce an artificial time-damping to the Navier-Stokes equations that can result in numerical steady solutions that are actually physically unstable. On the other hand, it can be advantageous to obtain a steady solution for a possibly unstable flow, e.g. a separated shear layer, which can then be analyzed for stability using analytical and/or computational techniques.

Bearing these considerations in mind, it will become clear that the Semidirect methods have significant potential for multidimensional flow calculations, and for other nonlinear steady-state problems in areas such as heat transfer, electrodynamics, grid generation, plasticity, etc.

8.2 Burgers Equation by Time Accurate Methods

As a 1-D prototype for multidimensional fluid dynamics, we consider the nonlinear (quasilinear) Burgers equation [1] for $u(x,t)$, usually written as

$$u_t = -u u_x + \frac{1}{R_e} u_{xx} \tag{8.2.1}$$

R_e is the Reynolds number, a physical parameter of the problem that indicates the relative size of viscous and inertia forces, with lower R_e meaning a more viscous or viscosity-dominated problem. An alternate normalization is obtained by multiplying Eq. (8.2.1) by R_e, giving

$$u_\tau = -R_e u u_x + u_{xx} \tag{8.2.2}$$

with $\tau = R_e/t$. The latter form is more appropriate for low R_e limits, and the former for high R_e limits. (Both are written here in "non-conservation" form [1].) Also, a linearized form of Burgers equation is useful. In this simple advection-diffusion model equation, the non-linear term u in Eq. (8.2.2) is replaced by a linear advection speed c, giving

$$u_\tau = -R_e \cdot c\, u_x + u_{xx} \tag{8.2.3}$$

As a prototype of the explicit time-dependent methods, we consider the forward-time, centered-space (FTCS) [1] method applied to (8.2.1),

$$\frac{u_i^{n+1} - u_i^n}{\Delta t} = -u_i^n \frac{u_{i+1}^n - u_{i-1}^n}{2\Delta x} + \frac{1}{R_e}\left[\frac{u_{i+1}^n - 2u_i^n + u_{i-1}^n}{\Delta x^2}\right] \tag{8.2.4}$$

For the linear equation (8.2.3), u_i^n above is replaced by c.

The performance of the FTCS is probably a *lower* bound on performance for time-dependent methods, yet it is representative of the performance of many explicit methods used for practical calculations.

Vastly superior performance is obtained with an implicit backward-time, centered space (BTCS) method. The derivatives of the right member of Eq. (8.2.4) are now evaluated at the advance time level $n+1$, giving an implicit method in which all the values u_i^{n+1} are solved simultaneously by a matrix inversion. A nonlinearly implicit method would evaluate the nonlinear advection term also at $n+1$. However, this would result in a nonlinear implicit equation for the u_i^{n+1} which would require an iterative solution within each time step. It is well recognized that this is an inefficient procedure, and a common approach is to lag the nonlinear terms in the time iteration scheme, giving

$$\frac{u_i^{n+1} - u_i^n}{\Delta t} = -u_i^n \frac{u_{i+1}^{n+1} - u_{i-1}^{n+1}}{2\Delta x} + \frac{1}{R_e} \frac{u_{i+1}^{n+1} - 2u_i^{n+1} + u_{i-1}^{n+1}}{\Delta x^2} \tag{8.2.5}$$

After multiplication by $R_e \Delta x^2 \Delta t$, the matrix equation for u_i^{n+1} is arranged as

$$\left[d + \frac{1}{2}C_i^n\right] u_{i-1}^{n+1} - (2d+1)u_i^{n+1} + \left[d - \frac{1}{2}C_i^n\right] u_{i+1}^{n+1} = -u_i^n \tag{8.2.6}$$

where the local Courant number is $C_i^n = \dfrac{u_i^n \Delta t}{\Delta x}$ and $d = \dfrac{\Delta t}{R_e \Delta x^2}$.

This equation is tridiagonal and may be readily inverted. For the linear equation, u_i^n is replaced by C.

There are many other time-dependent methods, of course, (e.g., see [1]) but the explicit FTCS and implicit BTCS seem to be representative of the extremes of poor and good performance for steady-state convergence.

Other iterative methods are ostensibly steady-state methods, but turn out, upon examination, to be time-like. For example, Richardson's method for elliptic equations begins with a steady-state equation, but it can be shown [1] to be algebraically identical to FTCS with $\Delta t = \Delta t_{crit}$. Most other point- and line-relaxation methods are time-like in their iterations, if not actually time-dependent, and their iterative convergence rates are represented by the FTCS and BTCS methods.

8.3 Basic Idea of Semidirect Methods

A remarkable point about these and similar time-dependent or time-like methods is that the algorithms are virtually identical for a *linear* problem as for a *nonlinear* problem. In fact, the methods will sometimes converge *more slowly* for a *linear* problem. *No information about the linearity is used to any advantage*, even though one could solve the linear steady-state problem directly, i.e., non-iteratively. In Semidirect methods, we take advantage of the ability to solve linear problems non-iteratively.

The basic idea of Semidirect methods is to assume a steady state, linearize the steady-state equation, solve this linear equation *directly* (i.e., non-iteratively), and then iterate to remove the nonlinearity.

We will demonstrate this idea with the most obvious linearization for the Burgers equation.

8.4 Burgers Equation by Picard Semidirect Iteration

The Picard iteration scheme is a simple Semidirect method for obtaining nonlinear solutions. It has been used as a method of obtaining a constructive proof of existence of solutions for nonlinear equations (e.g., see Weinberger [2]). We assume a steady-state, so $u_\tau = 0$ in the Burgers equation (8.2.2), and lag the nonlinear term in the iteration. The iteration counter is indicated by k, so we write

$$u_{xx}^k - R_e u^{k-1} u_x^k = 0 \tag{8.4.1}$$

This equation represents a sequence in k of linear problems; to emphasize this, we re-write it in terms of the linear operator L as

$$L^{k-1} u^k = 0 \tag{8.4.2}$$

where

$$L^{k-1} = \frac{\partial^2}{\partial x^2} - R_e u^{k-1} \frac{\partial}{\partial x} \tag{8.4.3}$$

Starting from some initial guess u^0, the sequence of successive approximations u^1, u^2, etc. is obtained from solutions of

$$L^0 u^1 = 0, \quad L^0 = \frac{\partial^2}{\partial x^2} - R_e u^0 \frac{\partial}{\partial x} \tag{8.4.4a}$$

$$L^1 u^2 = 0, \quad L^1 = \frac{\partial^2}{\partial x^2} - R_e u^1 \frac{\partial}{\partial x} \tag{8.4.4b}$$

$$L^2 u^3 = 0, \quad L^2 = \frac{\partial^2}{\partial x^2} - R_e u^2 \frac{\partial}{\partial x} \tag{8.4.4c}$$

Each linear problem has the following tridiagonal matrix form, obtained by using $0(\Delta x^2)$ centered differences in Eq. (8.4.1) and multiplying through by Δx^2.

$$\left[1 + \frac{1}{2} Rc_i^{k-1}\right] u_{i-1}^k - 2u_i^k + \left[1 - \frac{1}{2} Rc_i^{k-1}\right] u_{i+1}^k = 0 \tag{8.4.5}$$

where Rc_i is the local cell Reynolds number,

$$Rc_i^{k-1} = R_e \cdot \Delta x \cdot u_i^{k-1} \tag{8.4.6}$$

Again, each solution is obtained for u^k using the tridiagonal algorithm.

The performance is very good compared to the time-dependent method. For a stagnation-like flow with u(0) = 1 and u(1) = 0, for $R_e = 1$ and IL = 11 (i.e., 10 discrete cells), convergence to a tolerance of $\epsilon = 10^{-5}$ is obtained at six iterations or NIT = 6. This is comparable to the fully implicit BTCS method. However, the great advantage shows at IL = 101, for which convergence is *still* obtained at NIT = 6. For the higher value of $R_e = 10$ with IL = 101, convergence is obtained at NIT = 11 with Picard iteration with no relaxation. For comparison, the explicit FTCS time-dependent method *with optimum time step* requires 8852 iterations!

8.5 Further Discussion of the Picard Semidirect Iteration

The Picard Semidirect method for the Burgers equation has three clear advantages over the time-dependent methods. First and most obvious, it converges in fewer iterations than the time dependent methods, with no more work per iteration than the implicit BTCS time-step. Second, the convergence rate is not significantly affected by the problem size IL, which contrasts greatly to the explicit time-dependent methods. Third, it requires no search for an optimum Δt (or some analogous under-relaxation parameter in other time-like methods). This is a significant advantage, since the optimum Δt can be strongly dependent on Δx, Δy, R_e, and the problem geometry. (However, some Semidirect methods require an under-relaxation parameter, as we shall see.)

The iteration counter k is in some sense analogous to the time-level counter n of the time-dependent methods, and of course, the steady solution develops in the computer in real ("wall-clock") time as k increases. However, it is worth emphasizing that this Picard method is *not* really time-like. At each iteration level, a *steady-state solution* is obtained to a linearized problem. For example, if we make the crude initial guess of $u_i^o = 0$ at all $1 < i < IL$, the first iteration gives the steady-state solution to the zero-Reynolds number problem (called Stokes flow in fluid dynamics problems). This is in sharp contrast to time-like methods, in which the solution would develop in time even for $R_e = 0$. As k increases, the solution evolves *not* as a sequence of flow snapshots of time-developing flow, but more nearly as a sequence of steady-state nonlinear solutions, each at a higher R_e.

This description is not precisely correct, but it can be made so by "stacking" cases of increasing R_e at small increments, say $R_e = 0, 0.1, 0.2, \ldots$, etc. in a continuation method. The solutions for this R_e sequence differ so little that we would find each iteration produces an adequately converged solution, and a sequence of steady-state solutions is produced.

It is also worth noting that the iteration scheme Eq. (8.4.1) is described without recourse to the discretization method used. It could just as well apply to first-order spatial differences, or fourth-order, or a finite element formulation, or pseudo-spectral, or even to an analytical solution to the linear equation. In the latter case, the iteration scheme describes the analytical technique of a *regular perturbation method* based on a low R_e expansion.

Finally, note that each step of the Semidirect, non-time-like Picard method is identical to the implicit time-dependent BTCS method with $\Delta t \to \infty$. (This will not be the case for all Semidirect methods.)

8.6 Genesis of Semidirect Methods

Iteration schemes other than Picard iteration can be derived for the Burgers equation, and we will consider others shortly, but the Picard iteration serves well as a point of departure for multidimensional fluid dynamics methods. These differ from the Burgers equation qualitatively in the number of equations to be solved, in complexity of boundary conditions, and especially in dimensionality.

Dimensionality is the most significant obstacle. Whereas the $0(\Delta^2)$ differencing of the steady linearized Burgers equation gives rise to a tridiagonal matrix equation that is readily solved, the corresponding two-dimensional problem gives rise to a "block-tridiagonal" system that is not so easy to solve. The block-tridiagonal matrix is sparse and, for simple equations like the Poisson equation, is diagonally dominant. For moderate grid resolution, these are well suited for a solution by methods such as SOR (successive over-relaxation) and ADI (alternating direction-implicit), which are themselves time-like iterative methods [1]. Methods such as Gaussian elimination are applicable only to moderate 2-D problems because of round-off error accumulation, large computing time and excessive storage requirments (see Appendix C). Until 1965, iterative time-like methods were the only competitive methods for solving reasonably large systems of even the linear equations. Thus, Semidirect methods were not feasible.

In 1965, Hockney [3] published a direct method for solving a class of linear elliptic equations. Emphasis was on the constant-coefficient Poisson equation, one of the most common equations of mathematical physics. Other "fast Poisson solvers" soon appeared. In 1970, significant advances were made [4,5] in utilizing marching methods for direct solutions of variable-coefficient elliptic equations, including advection terms in the linear operation.

With fast, direct solvers available for linear elliptic equations, Semidirect solutions of nonlinear equations virtually suggest themselves, as exemplified by the following personal anecdote.

In early 1971, I completed the first draft of a book [1], *Computational Fluid Dynamics*, including a strong statement that time-dependent methods were the only recommended approach for steady-state Navier-Stokes problems. On a second reading of the manuscript, I began to wonder if this were really true. As I followed the historical development of time-like methods for linear equations, followed by time-like methods for non-linear equations, followed by direct methods for linear equations, the next step of Semidirect methods for nonlinear equations was obvious. By the time of the first printing of the book in December of 1972, I had changed my mind considerably and could refer to encouraging results from preliminary work on what we now call "Semidirect methods."

The most general solution techniques for the linear equations are the marching methods described earlier, that allow direct solution of variable-coefficient, non-separable elliptic equations including first- and cross-derivative terms. For this reason, and because my own experience is in this area, marching methods for the linear solvers will be emphasized in this book. Some of the Semidirect methods can also utilize other, more limited linear solvers.

In the previous sections, we began the presentation of Semidirect methods for fluid dynamics problems with the Burgers equation, a one-dimensional nonlinear equation that necessitates only the 1-D linear solver for tridiagonal equations. In Chapters 1-8, we presented fast multidimensional linear PDE solvers, so we are in a position to solve 2-D nonlinear fluid dynamics. However, there are complications besides dimensionality. One complication is the nature of the marching methods, with a significant difference in computer times between initialization and repeat solutions. The other complication is what primarily distinguishes a 2-D fluid dynamics problem from just a 2-D variant of Burgers equation: the complication of boundary conditions, which will have a major impact.

8.7 NOS Method

The first Semidirect methods we consider are the NOS and LAD methods (see Roache [6-10]). The governing equations used are the vorticity transport equation, the Poisson equation for stream function, and the relations between Ψ and the velocity components u and v. Subscripts indicate partial derivatives, as in $\zeta_t = \partial\zeta/\partial t$.

$$\zeta_t = -R_e \nabla \cdot \nabla\zeta + \nabla^2\zeta \tag{8.7.1a}$$

$$\nabla^2\Psi = \zeta \tag{8.7.1b}$$

$$V = u\vec{i} + v\vec{j} = \Psi_y\vec{i} - \Psi_x\vec{j} \tag{8.7.1c}$$

R_e is the Reynolds number. The interior equations (8.7.1) are supplemented by boundary conditions proper to a particular problem.

The first method is a Picard-type method, like the one used previously for the Burgers equation. It is similar to an n-th order Oseen approximation [11] solved numerically. We refer to it as the Numerical Oseen method and, following custom, we abbreviate it with the 3-letter acronym NOS.

After an initial guess on Ψ at all points, the iterative cycle begins with the boundary evaluation of ζ. Then the steady state form of Eq. (8.7.1a) is solved for the k-th iteration, using $0(\Delta x^2)$ centered differences, as

$$0 = -R_e \nabla \cdot \left(\nabla^{k-1}\zeta^k\right) + \nabla^2\zeta^k \tag{8.7.2}$$

The direct solution of Eq. (8.7.2) is obtained using the marching method of Chapter 2. Then the Poisson equation (8.7.1b) is solved for Ψ (by a direct method) and the cycle is repeated. Again, it is worth emphasizing that this method is not time-like; at each iteration, the linearized vorticity transport equation is solved exactly for the steady-state solution.

This NOS method requires the solution of a second-order linear finite-difference equation with first-order terms. Since the coefficients of these terms (u and v) vary with each iteration, the marching solution must be initiated each time, i.e., a new influence coefficient matrix must be generated and "inverted." This is in contrast to the Poisson solution, which need be initiated only once. Also, the applicability is limited by the restrictions of the marching method.

8.8 LAD Method

A second method that avoids these shortcomings is obtained by replacing the vorticity iteration of NOS by

$$\nabla^2 \zeta^k = F^{k-1} \equiv R_e \nabla \cdot \left(V^{k-1} \zeta^{k-1}\right) \tag{8.8.1}$$

In this method, only the Laplacian term of the vorticity transport equation drives the iterations toward a solution; the method is accordingly described as the Laplacian Driver method and abbreviated as LAD. Other aspects are the same as NOS. This method may use a marching method or other direct Poisson solvers such as Hockney's method[3] or Buneman's method [12].

8.9 Performance of NOS and LAD on the Driven Cavity Problem

These methods were tested [6-8] for the geometrically simple driven cavity problem, as shown in Figure 8.9.1. The continuum boundary conditions for this problem are simple and unambiguous, compared to flow-through problems. On all boundaries, the non-slip conditions prevail, giving $\Psi = 0$ and allowing the evaluation of ζ by any of three equations, written here for the moving 'lid' at $j = J$. (For derivations and references, see [1].)

$$\zeta_{i,j} = \frac{2\left(\Psi_{i,J-1} - \Psi_{i,J} + \Delta y \cdot U\right)}{\Delta y^2} + 0\left(\Delta y\right) \tag{8.9.1a}$$

$$\zeta_{i,j} = \frac{3\left(\Psi_{i,J-1} - \Psi_{i,J} + \Delta y \cdot U\right)}{\Delta y^2} - \frac{1}{2}\zeta_{i,J-1} + 0\left(\Delta y^2\right) \tag{8.9.1b}$$

$$\zeta_{i,J} = \frac{-\Psi_{i,J-2} + 8\Psi_{i,J-1} - 7\Psi_{i,J} + 6\Delta y \cdot U}{2\Delta y^2} + 0\left(\Delta y^2\right) \tag{8.9.1c}$$

Equation (8.9.1a) is a first-order method. The method (8.9.1b) is due to Woods. Jensen's method Eq. (8.9.1c) is obtained by passing a cubic equation for Ψ through the wall point and two neighboring points. Briley [13] has shown in time-dependent calculations using Eq. (10.2.4c) that the stability and accuracy of the total solution is enhanced if the velocities near the wall are also evaluated in a form compatible with the assumed cubic form of Ψ, so u near the lid is evaluated by

$$u_{i,J-1} = \frac{-\Psi_{i,J-2} - 4\Psi_{i,J-1} + 5\Psi_{i,J}}{4\Delta y} - \frac{1}{2}u_{i,J}. \tag{8.9.2}$$

Similar equations apply at the other walls. Since R_e is based on the lid velocity, $u_{i,J} \equiv U = 1$.

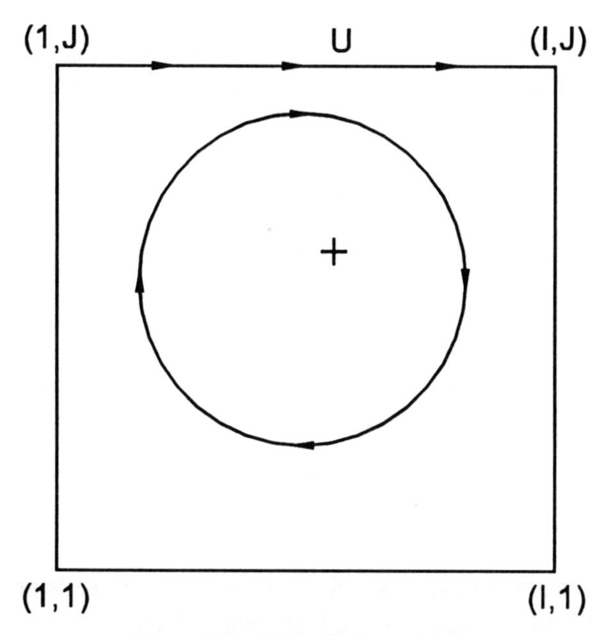

Figure 8.9.1. Driven Cavity problem.

Unlike the situation for the 1-D Burgers equation, both NOS and LAD methods were *unstable* at all R_e tested, including $R_e = 0$. In order to achieve stability, it is necessary to under-relax the boundary evaluations of vorticity (ζ_b) as in

$$\zeta_b^{k+1} = f \cdot r + \zeta_b^k (1 - r) \tag{8.9.3}$$

where f is a function indicated in Eq. (8.9.1) and r is the under-relaxation parameter, $0 < r < 1$. It is not necessary to under-relax separately the interior values of ζ or Ψ. In the work of Ehrlich [14], the biharmonic equation was solved by coupling the solution of two Poisson equations at the boundary, equivalent to the present problem with $R_e = 0$. All the interior values of ζ or Ψ were under-relaxed. This effectively under-relaxes ζ_b through Eq. (8.9.1a), but the experiments in [6-8] indicate that the under-relaxation at other points is unnecessary.

For comparison, the FTCS time-dependent method was also tested. The complete tests varied the method (NOS, LAD, and FTCS), the wall vorticity equation (8.9.1a, b, c), the under-relaxation parameter ($0 < r \leq 1$) and the flow parameter R_e($0 \leq R_e \leq 100$). Initial conditions were the no-flow conditions of $\Psi = 0$ and $\zeta = 0$ everywhere. All cases were carried out to 111 iterations, unless strong instability occurred. All field variables and several diagnostic functionals were examined subjectively for iteration convergence. In some cases (first-order wall ζ, low R_e and near-optimum r) iteration convergence was unequivocal, with floating-point zeroes for the change in quantities indicating convergence to the single-precision accuracy of the computer ($\approx$ 14 decimal significant figures).

To quantify the comparisons in those tests, the functional $\Sigma_{ij}|\zeta_{ij}|$ was evaluated, where the summation extends *only over the internal points* $2 \leq i \leq I-1$, $2 \leq j \leq J-1$. Then

$$DQ^{k+1} = \Sigma_{ij}|\zeta_{ij}^{k+1}| - \Sigma_{ij}|\zeta_{ij}^{k}| \tag{8.9.4a}$$

was tested, and the number of iterations NIT required to obtain

$$|DQ|/r < 10^{-4} \tag{8.9.4b}$$

was tracked. This was approximately equivalent to requiring a normalized value of

$$|DQ|/\left(r|\Sigma_{ij}\zeta_{ij}^{k}|\right) \prec 10^{-6}. \tag{8.9.4c}$$

Since DQ may be made arbitrarily small by choosing r sufficiently small, clearly DQ/r is a better index of convergence than just DQ. For a discussion of convergence criteria, see [1,6-10,15]. (The reason for evaluating $\Sigma_{ij}|\zeta_{ij}|$ only over internal points will be clear shortly.)

The results for $R_e = 0$ and 20 using the first-order wall ζ evaluation, equation (8.9.1a), are shown in Fig. 8.9.2a. At $R_e = 0$, the NOS and LAD methods are identical. Also shown are results for the explicit time-dependent FTCS method. (The abscissa for the FTCS method is $\Delta t/\Delta t_{crit}$; at $R_e = 20$, the Δt_{crit} is controlled by a diffusion limitation rather than by a local Courant number limitation.)

For the driven cavity problem, the NOS and LAD methods are not significantly different in their convergence properties, and both exhibit an expected optimum value of r denoted by r_{op}. Beyond the vaguely-defined r_{op}, the convergence rate deteriorates rapidly (more so for LAD than NOS) and the methods then become unstable. This behavior is similar to the FTCS dependence on $\Delta t/\Delta t_{crit}$.

The results for other wall ζ evaluations of R_e are shown in Figs. 8.9.2b, c, which correspond to the use of equations (8.9.1b,c), respectively, for wall ζ. At $R_e = 0$, $r_{op} \simeq 1/4$ for the first-order wall ζ evaluation and about *1/5* for the second-order equations. As R_e increases, r_o increases slightly for NOS. For $R_e = 100$, convergence is not obtained. The use of the second-order methods for wall ζ causes faster instability for $r > r_o$, and generally slows convergence slightly for all r.

It is worth emphasizing that when iteration convergence did occur, all three methods (NOS, LAD, and FTCS) did converge to the same answer, independent of r and Δt.

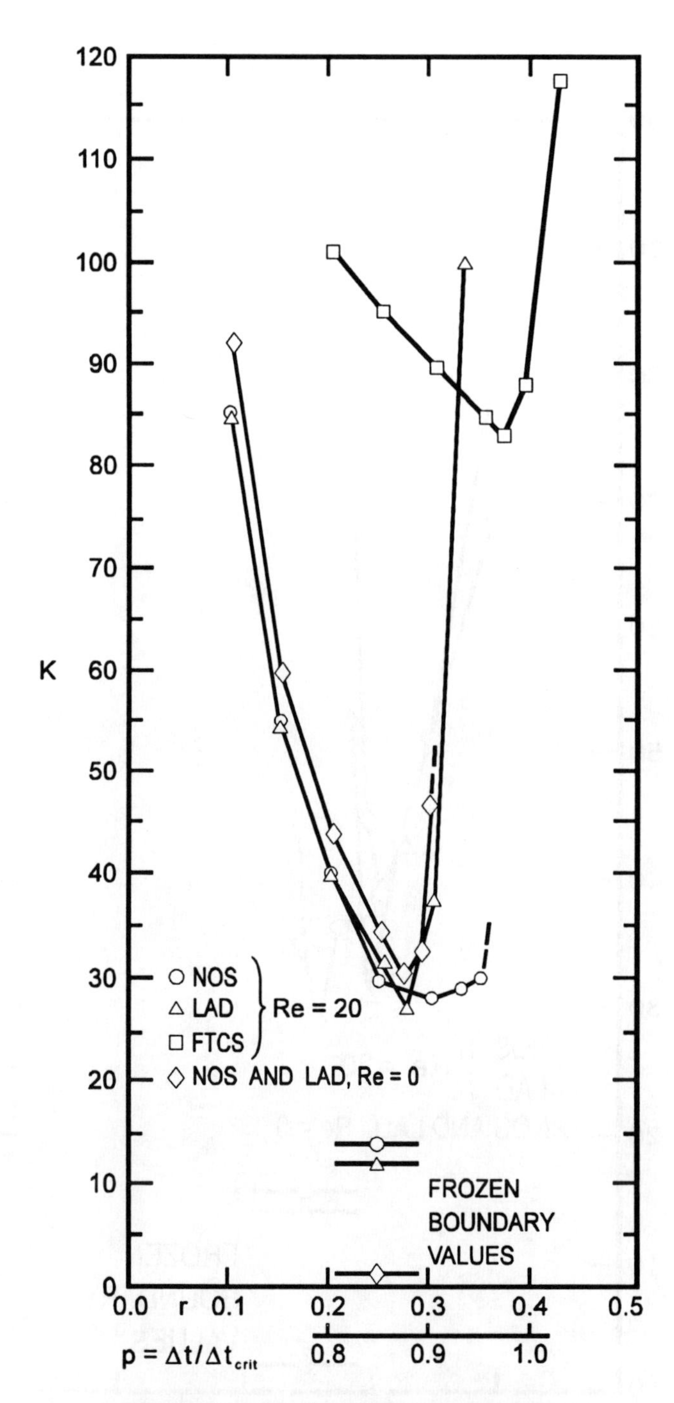

Figure 8.9.2a. Equation 8.9.1a for wall vorticity.
(The p-scale refers to the time-dependent FTCS method.)

Figure 8.9.2. Convergence results for a driven cavity problem, Re = 20, $\Delta x = \Delta y = 1/10$; r is the under-relaxation parameter. K is the number of iterations required to satisfy $DQ^K / r < 0.0001$.

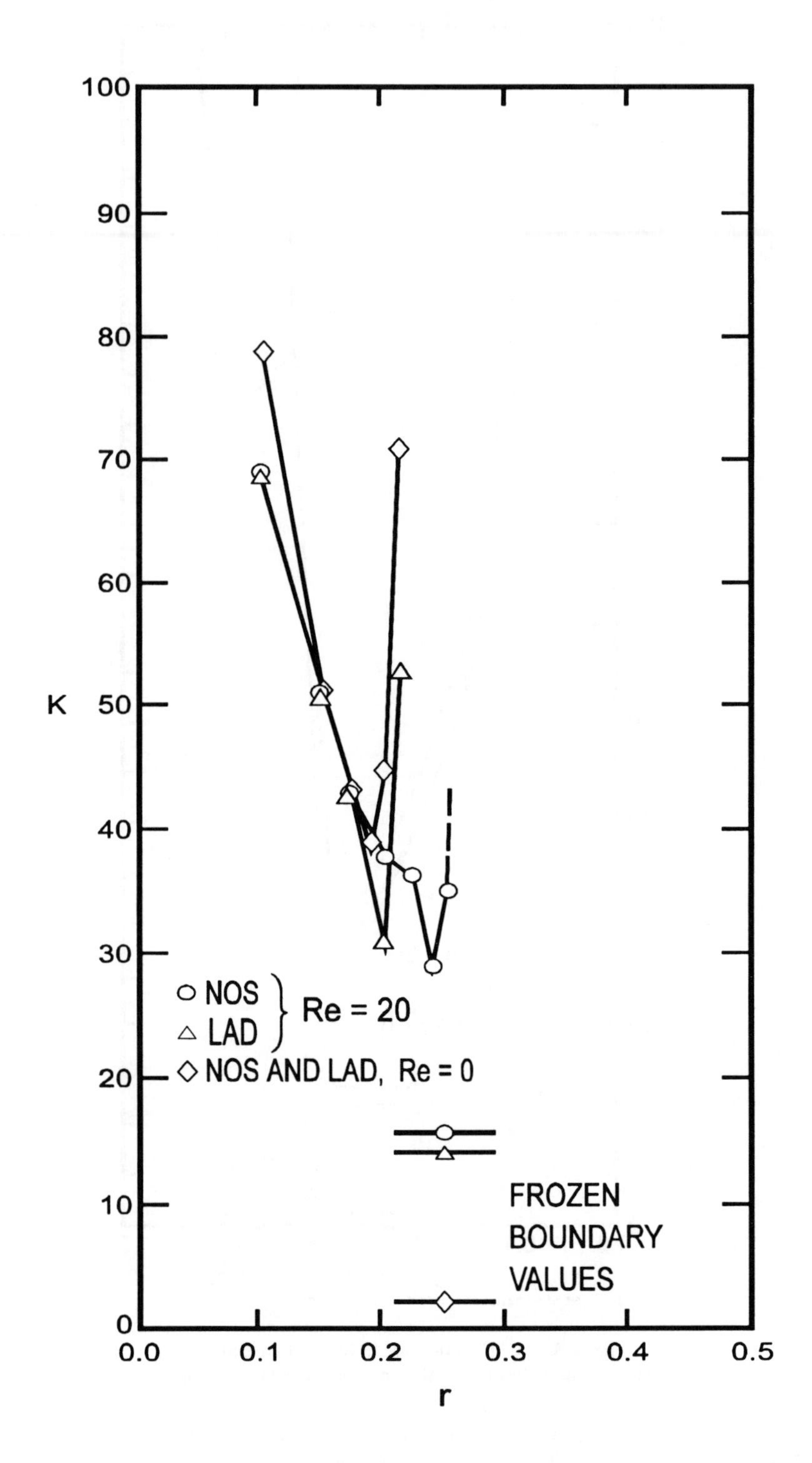

Figure 8.9.2b. Equation 8.9.1b for wall vorticity.

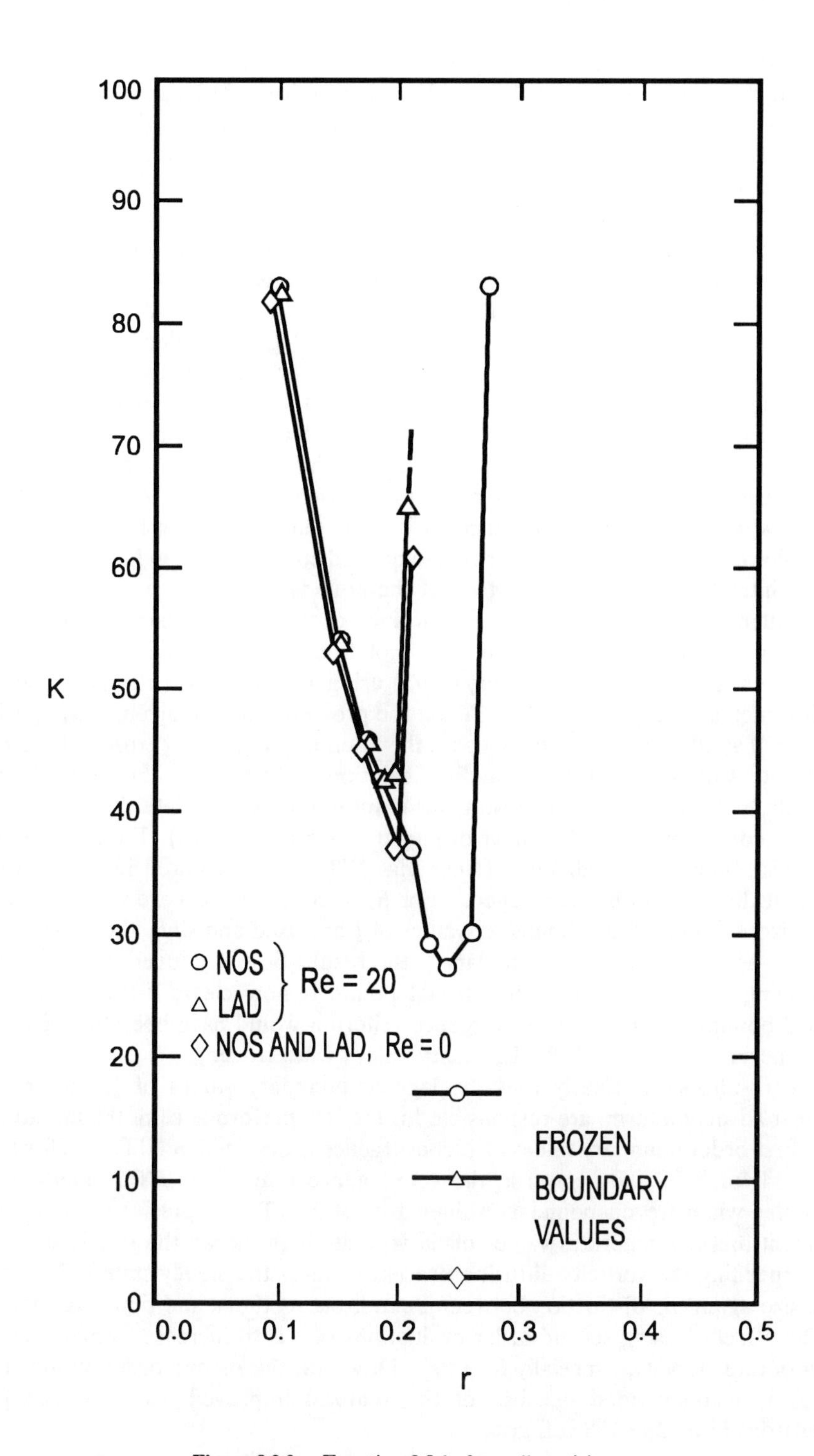

Figure 8.9.2c. Equation 8.9.1c for wall vorticity.

8.10 Relative Importance of Lagging Boundary Conditions

From the results in Fig. 8.9.2, it is clear that both the NOS and LAD methods are superior to the explicit time-dependent FTCS method in regard to the number of iterations required to meet the convergence criterion used. However, the improvement is not at all as large as might be expected from the experiments on the Burgers equation. It is also notable that the NOS method, which includes advance information in the advection term, was not clearly faster (in the number of iterations) than LAD. Both of these observations suggested the experiment described below.

There are two soure of coupling between the Ψ and ζ interior equations. The first and most obvious source is the lagging velocity field u, v = $fcn(\Psi)$ in the advection term of the NOS method, or the velocity field and ζ field in the lagging advection term, F, of the LAD method. The second source is the boundary condition on ζ, equation (8.9.1), through which ζ_b depends on internal values of Ψ. Because of the small difference between the behavior of NOS and LAD (which treat the nonlinear advection term differently) and between cases with $R_e > 0$ compared to $R_e = 0$ (which has no advection term), it is clear that the lagging boundary values of ζ are a major contributor to the slower-than-expected convergence. In fact, the requirement to lag ζ_b introduces a *time-like behavior* at the boundaries.

To quantitatively determine the importance of the lagging boundary conditions, the following procedure was followed [6-8]. A solution for a particular R_e was determined, starting with $\Psi = 0$ and $\zeta = 0$ everywhere, using $r \simeq r_o$ and a stringent convergence criterion such as $|DQ|/r < 10^{-10}$. Then the problem was run again, starting with $\Psi = 0$ and $\zeta = 0$ at all internal points, but with the boundary values of ζ frozen at their correct steady-state values obtained in the first solution. (Obviously, this procedure is not applicable to solving flow problems; it was employed only to isolate the effect of lagging boundary conditions from the effect of lagging advection terms.) For $R_e = 0$, the final solution is obtained immediately. (Note that NIT = 2 is required just to methodically verify that the solution has converged.) For $R_e = 20$, the results are plotted in Fig. 8.9.2 as a horizontal line. (Since boundary values of ζ are fixed and since the under-relaxation factor r is applied only to the boundary ζ, the results do not depend on r.) The reason for defining $\Sigma_{ij} |\zeta_{ij}|$ only over internal points is now clear; if the summation had included boundary points, the convergence criterion would have been biased in favor of these experiments for which the boundary values were frozen.

The results show clearly that the lagging boundary values of ζ, rather than the nonlinear advection term, are responsible for the low performance of the methods. With frozen first-order boundary values of ζ, convergence is obtained at NIT = 12 for LAD and NIT = 14 for NOS. By contrast, the convergence rate of the FTCS method actually *deteriorates* when frozen boundary values are used. (The impulsive start of the time-dependent method generates values of lid vorticity higher than the steady-state "frozen" values, speeding the vorticity diffusion process towards the steady state.) This serves to emphasize again the distinction between Semidirect methods and time-like methods.

The effect of using second-order evaluations of ζ is to increase formal accuracy but to deteriorate stability, especially for $r > r_o$. However, the second-order Woods equation (8.9.1b) is recommended because of the realized improved accuracy; see [8,9] for comparisons in a 128×128 cell grid.

The relative importance of the lagging boundary conditions was further demonstrated by calculations of a driven cavity with various (unstable) symmetries, giving $\zeta = 0$ along some boundaries [8].

8.11 Performance of NOS and LAD on a Flow-Through Problem

A driven cavity problem is diffusion dominated. (The R_e is based on the lid speed, but the speed at interior points is much lower.) It would appear that the NOS method, which includes advance information in the advection term, might require significantly less iterations to converge than LAD for a *flow-through problem* , wherein the internal speeds are larger.

A developing channel flow shown in Fig. 8.11.1 served for comparison [8,9]. The inflow boundaries Ψ and ζ are fixed with a linear u profile and $v = 0$, i.e., Couette flow. The maximum value $u = 2$ was chosen so that R_e is based on the bulk mixing velocity rather than the maximum velocity. Thus, $\Psi(y) = y^2$ and $\zeta(y) = 2$ at inflow. At $y = 1$, the upper boundary of the symmetric lower-half of the problem, we have $\Psi = 1$ and $\zeta = 0$. Along the lower boundary, no-slip wall conditions apply, with $\Psi = 0$ and ζ given by Eq. (8.9.1). The outflow boundary at $i = I$ was set by methods commonly used in time-dependent computations [1], using $0(\Delta^2)$ finite-difference analogs of

$$\partial^2 \Psi / \partial x^2 = 0 \tag{8.11.1a}$$

$$\partial \zeta / \partial x = 0 \tag{8.11.1b}$$

The first condition was also set at $i = I$, allowing the solution of an ordinary finite difference equation for Ψ. The cell aspect ratio is $\Delta x / \Delta y = 10$. Convergence results for $R_e = 1$ and $R_e = 50$ are presented in Fig. 8.11.2. As expected, NOS performs significantly better than LAD in the number of iterations required. Also, the presence of three boundaries on which the steady-state conditions on values of Ψ and ζ are known significantly speed convergence, compared with the driven cavity problem. (See [8,9] for effects of various wall vorticity equations.) At high R_e , LAD did not converge, even for r as low as 0.05. For R_e from 1 to 10, NOS required less iterations than LAD by a factor of about 4.5 with second-order wall ζ and about 3 to 4.5 with first-order wall ζ. (LAD similarly fails to converge for high R_e Burgers equations, indicating that it is not just a problem with fluid dynamics boundary conditions.)

Most of the time-dependent FTCS results were off the scale of Fig. 8.11.2. The best performance was at $R_e = 50$ using equation (8.9.1b) for wall vorticity; for the best time step, this case converged at NIT = 54 which is more than 3 times larger than the NOS value. For $R_e = 1$, the best performance of FTCS was NIT = 106 which is more than 2 times larger than the LAD value and more than 10 times larger than the NOS value. Note that NOS gave converged answers in 9 or 10 iterations, whereas the explicit time-dependent FTCS method requires 9 iterations just for *any* information from the inflow boundary to be advected through the mesh.

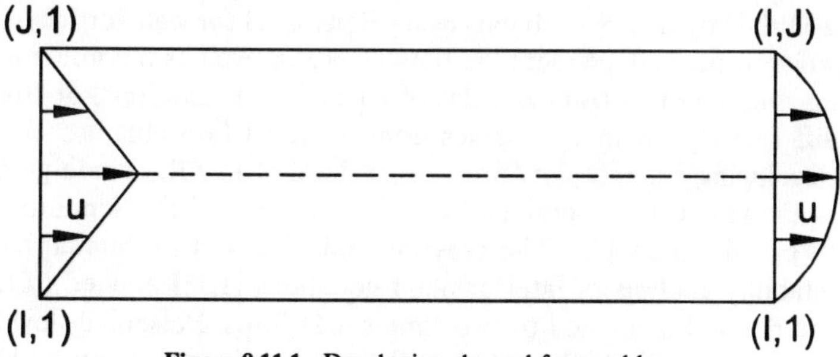

Figure 8.11.1. Developing channel fow problem.

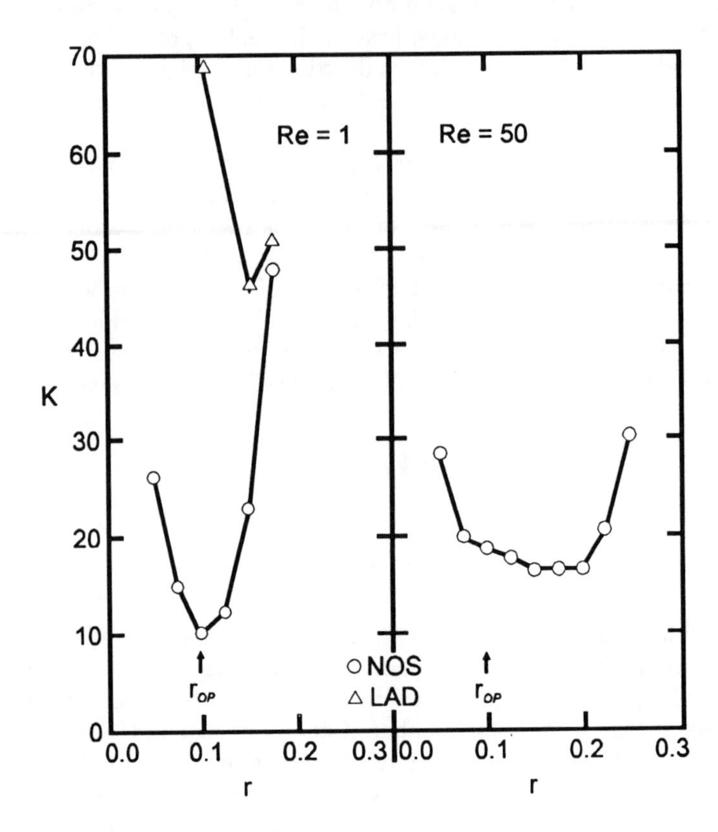

Figure 8.11.2. Convergence results for developing channel flow, $\Delta y = 1/10$, $\Delta x = 1$. Equation 8.9.1b for wall vorticity.

As R_e increases, the curve of K versus r becomes flatter, i.e., the convergence rate becomes less sensitive to r near r_{op} (contrasted to the sensitivity of SOR relaxation near its optimum over-relaxation parameter [1]). This is helpful because we can only predict optimum r_{op} at low R_e , as in the next section.

8.12 Optimum Relaxation Factor and Convergence for Large Problems

An advantage of Semidirect methods in 1-D is that the number of iterations required to reach convergence is approximately independent of problem size (for small Δx). However, in 2-D fluid dynamics problems using Eqns. 8.9.1 for wall vorticity, the optimum relaxation factor r_o is mesh dependent. Estimates of r_o as well as the minimum r required for stability may be obtained from a series of coarse-mesh solutions, or from fine-mesh solutions at different R_e , or in some cases from related 1-D problems.

For the developing channel flow problem, at least at low R_e , some guidance on the stability and optimum r can be obtained from consideration of the Semidirect solution to a simple 1-D Poiseuille flow [6]. The present stability analysis is similar to the discrete perturbation stability analysis of interior-point equations [1,16] applied to the boundary equations. The method is applied to two simple 1-D flows, Poiseuille flow and Couette flow. For the simple Poiseuille flow, the nonlinear advection terms of the Navier-Stokes equations. The method is applied to two simple 1-D flows, Poiseuille flow and Couette flow. For the simple Poiseuille flow, the nonlinear advection terms of the Navier-Stokes equations are absent; as in the case of $R_e = 0$ flows, the NOS and LAD methods are then indistinguishable. The interior-point equations reduce to

$$\frac{d^2\Psi}{dy^2} = \zeta, \quad \frac{d^2\zeta}{dy^2} = 0 \tag{8.12.1}$$

The approach of the discrete perturbation staility analysis is to first assume that a steady-state solution ζ exists. Then an arbitrary discrete error ϵ is introduced into the boundary calculation. (This error ϵ, not necessarily small, may be thought of as being introduced by machine round-off error, for example.) Then the effect of the error is followed as the iteration proceeds. Stability is indicated if the error diminishes as the iterations proceeds.

It may be verified that the analytical solution of the finite-difference equations for the Poiseuille flow, using second-order centered differences, is given as follows. The subscript j runs from $j = J$ at the center of the flow gap, where $u = 1$.

$$\Psi_1 = 0, \quad \Psi_2 = \frac{\Psi_J}{J-1} - \zeta_1 \Delta y^2 \varphi(J) \tag{8.12.2}$$

$$\Psi_j = (j-1)\Psi_2 + \frac{\zeta_1 \Delta y^2}{J-1} \sum_{\ell=2}^{j-1} (j-\ell)(J-\ell) \tag{8.12.3}$$

$$\Psi_J = (J-1)\Psi_2 + \zeta_1 \Delta y^2 (J-1)\varphi(J) \tag{8.12.4}$$

$$\zeta_j = \zeta_1 \left[\frac{J-j}{J-1}\right] \tag{8.12.5}$$

where $J = \max j$ and

$$\varphi(J) = \frac{1}{(J-1)^2} \sum_{k=1}^{J-2} k^2 \tag{8.12.6}$$

$$= \frac{(J-2)}{(J-1)} \frac{(J-3/2)}{3} \tag{8.12.7}$$

$$\simeq \frac{J}{3} \; for \; J \gg 1 \tag{8.12.8}$$

The wall vorticity value ζ_1 depends on the method used to evaluate wall vorticity as a function of the interior-point values. As an example, consider only the first-order method from Eq. 8.9.1a with wall velocity $U = 0$.

$$\zeta_1^a = \frac{2(\Psi_2 - \Psi_1)}{\Delta y^2} \tag{8.12.9}$$

In the total Poiseuille flow solution, this gives a solution from Eq. (8.12.4) of

$$\zeta_1^a = \frac{2\Psi_J}{\Delta y\,[1 + 2\varphi(J)]} \tag{8.12.10}$$

These results for ζ_1 are included here for the sake of completeness, although they will not be needed for the stability analysis.

Consider a 3-step cycle in the iteration procedure:

(1) ζ_1^{k+1} is solved by $\zeta_1^{k+1} = fcn(\Psi_2^k, \Psi_3^k, \zeta_2^k)$ where the boundary *fcn* is given by Eq. (8.12.9);

(2) ζ_j^{k+1} for all $1 < j < J$ is solved by $\delta^2\zeta / \delta y^2 = 0$ with ζ_1^{k+1} fixed by step (1) and $\zeta_J^{k+1} = 0$, from Eq. (8.12.5)

(3) Ψ_j^{k+1} for all $1 < j < J$ is solved by $\delta^2\Psi / \delta y^2 = \zeta^{k+1}$ through the application of Eq. (8.12.2-4).

Assume a steady-state solution, ζ (without superscript). At any point in the iteration cycle, the boundary value ζ_1 is perturbed by ϵ.

$$\zeta_1 \leftarrow \zeta_1 + \epsilon \tag{8.12.11}$$

It is readily seen that no other effect is produced until step (2), so that the perturbation may be applied after step (1) without loss of generality. The effect of the perturbation then propagates through the iteration cycle as follows.

$$(1) \quad \zeta_1^{k+1} = \zeta_1 \quad (\textit{steady-state solution}) \tag{8.12.12}$$

$$\ldots\textit{Perturbation:} \quad \zeta_1^{k+1} = \zeta_1 + \epsilon \ldots \tag{8.12.13}$$

$$\begin{aligned} (2) \quad \zeta_2^{k+1} &= \zeta_1^{k+1}\frac{J-2}{J-1} \\ \zeta_3^{k+1} &= \ldots \end{aligned} \tag{8.12.14}$$

$$\begin{aligned} (3) \quad \Psi_2^{k+1} &= \frac{\Psi_J}{J-1} - \zeta_1^{k+1}\Delta y^2\,\varphi(J) \\ \Psi_3^{k+1} &= 2\Psi_2^{k+1} + \zeta_2^{k+1}\Delta y^2 \\ &= 2\Psi_2^{k+1} + \frac{J-2}{J-1}\zeta_1^{k+1}\Delta y^2 \\ \Psi_4^{k+1} &= \ldots \end{aligned} \tag{8.12.15}$$

$$(4) \quad \zeta_1^{k+2} = fcn\left(\Psi_2^{k+1}, \Psi_3^{k+1}, \zeta_2^{k+1}\right) \tag{8.12.16}$$

Define f by

$$f = \frac{\zeta_1^{k+2} - \zeta_1}{\epsilon} \tag{8.12.17}$$

For stability, it is required that

$$|f| \leq 1 \tag{8.12.18}$$

First consider the first-order wall vorticity method with no under-relaxation. Equation (8.12.15) gives

$$\Psi_2^{k+1} = \frac{\Psi_J}{j-1} - (\zeta_1 + \epsilon)\Delta y^2 \varphi(J) \tag{8.12.19}$$

Comparing this to Eq. (8.12.2),

$$\Psi_2^{k+1} = \Psi_2 - \epsilon \Delta y^2 \varphi(J) \tag{8.12.20}$$

where Ψ_2 is the steady-state solution. Then Eq. (8.12.16) gives, using the first-order method Eq. (8.12.9),

$$\zeta_1^{k+2} = \frac{2\Psi_2^{k+1}}{\Delta y^2} = \frac{2\Psi_2}{\Delta y^2} - 2\epsilon\, \varphi(J). \tag{8.12.21}$$

Comparing this to Eq. (8.12.9),

$$\zeta_1^{k+2} = \zeta_1 - 2\epsilon\, \varphi(J) \tag{8.12.22}$$

Then by Eq. (8.12.17),

$$f \equiv \frac{\zeta_1^{k+2} - \zeta_1}{\epsilon} = -2\, \varphi(J). \tag{8.12.23}$$

From Eq. (8.12.6) we see that even for J as small as $J = 3$, we have $\varphi(3) = 2$, which gives $f = -2$. By Eq. (8.12.18) at interior points, this is unstable.

To achieve stability, we introduce an under-relaxation parameter r into the boundary evaluation of ζ_1, replacing iteration step (4), Eq. (8.12.16), by

$$\zeta_1^{k+2} = r \cdot fcn\left(\Psi_2^{k+1}, \Psi_3^{k+1}, \zeta_2^{k+1}\right) + (1-r) \cdot \zeta_1^{k+1}. \tag{8.12.24}$$

Then Eq. (8.12.21) is replaced by

$$\zeta_1^{k+2} = r \cdot \frac{2\Psi_2^{k+1}}{\Delta y^2} + (1-r)\,\zeta_1^{k+1}$$
$$= r\left[\frac{2\Psi_2}{\Delta y^2} - 2\epsilon\,\varphi(J)\right] + (1-r)\left(\zeta_1 + \epsilon\right). \tag{8.12.25}$$

Using Eq. (8.12.9) gives

$$\zeta_1^{k+2} = \zeta_1 - 2r\epsilon\,\varphi(J) + \epsilon(1-r). \tag{8.12.26}$$

Then by Eq. (8.12.17),

$$f = -2r\,\varphi(J) + 1 - r \tag{8.12.27}$$

For $r = 1$, we obtain $f = -2\,\varphi(J)$ as in Eq. (8.12.23). For $r = 0$, we obtain $f = +1$ or no iteration at all. The borderline for oscillatory stability occurs at $f = -1$. Using this value and solving Eq. (8.12.27) gives

$$r_{\text{lim}} = \frac{2}{1 + 2\,\varphi(J)}. \tag{8.12.28}$$

For $r < r_{\text{lim}}$, the method is stable. If instead, we set $f = 0$, from Eq. (8.12.17) we obtain $\zeta_1^{k+2} = \zeta_1$, the steady-state value. That is, for this simple problem, the final steady-state answer is reached in 1 iteration. This is clearly the *optimum* value r_{op} for this Poiseuille flow, $r_{op,P}$.

Setting $f = 0$ in Eq. (8.12.27) gives

$$r_{op,P}^{a} = \frac{1}{1 + 2\,\varphi(J)} \tag{8.12.29}$$

where the "a" superscript refers to the use of the first-order equation like Eq. (8.12.9) for ζ_1. Using the approximation Eq. (8.12.8) for $J \gg 1$ gives

$$r_{op,P}^{a} \simeq \frac{1}{1 + 2\left[\dfrac{J}{3}\right]} \simeq \frac{3}{2J} \tag{8.12.30}$$

Following similar steps [6], optimum values of r were predicted for both Couette Flow, $r_{op,\,C}$ and Poisseuille flow, $r_{op,\,P}$ considering all three forms Eq. (8.9.1) of wall vorticity. The summarized results are presented in Table 8.12.1. (The value of r_{op} for Poiseuille flow, $r_{op,\,P}$, was shown in Fig. 8.11.2).

For low R_e, the asymptotic Poiseuille solution develops quickly; $r_{op,\,P}$ is appropriate and accurately predicts the experimentally observed r_{op} for the NOS method at $R_e = 1$. For high R_e, the inflow Couette flow profile persists and the r_{op} for Couette flow, $r_{op,\,C}$, would seem to be appropriate. However, $r_{op,\,C}$ (not shown in Fig. 8.11.2) does *not* accurately indicate the experimentally observed r_{op} for $R_e = 50$, nor does it even indicate the correct trend. That is, $r_{op,\,C} < r_{op,\,P}$, but the experimentally observed r_{op} actually increases as R_e increases. The reason is apparently because the centerline value of vorticity for this 2-D problem is $\zeta_{i,J} = 0$, rather than the Couette-flow value of $\zeta_{i,J} = 2$.

	Wall vorticity method	r_{op}	r_{op} for J»1	r_{op} for J=11	r_{op} for J=101
Poiseuille flow	First-order	$\frac{1}{1+2\varphi(J)}$	$\frac{3}{2J}$	0·1493	0·01500
		$\varphi(J)=\left(\frac{J-2}{J-1}\right)\left(\frac{J-3/2}{3}\right)$			
	Second-order	$\frac{1}{J-1}$		0·1	0·01
Couette flow	First-order	$\frac{3}{5}\left(\frac{1}{J-1}\right)$		0·06	0·006
	Second-order	$\frac{2}{5}\left(\frac{1}{J-1}\right)$		0·04	0·004

Table 8.12.1. Optimum semidirect relaxation factors r on wall vorticity for Poiseuille Flow and Couette Flow with constant Δx, Δy. The stability limit (maximum) on r is given by 2 X optimum r shown in the table.

The value $r_{op,\,C}$ would be appropriate for a low R_e problem of developing Couette flow.

In a more general approach, inverted Aitken extrapolation was used [10] to estimate optimum r from the early iterative behavior. After three iterations, ζ_w^∞ is estimated using Aitken extrapolation, and r_{op} is approximated as the r that would have given ζ_w^∞ on the third iteration.

Ehrlich's [14] Semidirect solution of a linear ($R_e = 0$) problem using a somewhat different under-relaxation scheme showed the same trend. Under-relaxation of ψ and ζ was used at interior points, effectively relaxing ζ_b; note also that his w_{op}^2 is $\propto$ r. His analysis is consistent with the prediction $r_{op} \propto 1/\Delta$. He also predicted that the number of iterations required to reach convergence would vary as $\sqrt{1/\Delta x}$, but the analogy may not hold in the present iteration scheme as applied to the coupling of linear equations.

These considerations lead in Section 8.15 to an alternate formulation of wall vorticity that does not show strong dependence of r_{op} on mesh size.

8.13 Choice Between LAD and NOS

The LAD and NOS methods differ not only in the number of iterations required to meet a convergence requirement but also in the computation time per iteration. Since LAD involved only the solution of a Poisson equation, it is faster than NOS (which requires the generation and 'inversion' of a new influence coefficient matrix at each iteration) beyond the first iteration. This difference is especially important in large problems, in which the time to re-initiate the elliptic solver becomes large.

For the flow-through problem of developing channel flow, NOS required significantly fewer iterations to converge than LAD. For very small problems such as the 10×10 cell problem referred to earlier, NOS is also faster in computation time. But the faster computation speed of LAD per iteration will off-set the greater number of iterations, and LAD may be preferred over NOS for reasonably sized problems. (See, however, Chapter 4 on the banding approximation, which make NOS more than competitive.) Note also that LAD allows problems with different R_e (but in the same mesh with the same class of boundary conditions) to be 'stacked', i.e., a continuation method), with only one initiation of the Poisson solver for the entire stack.

The comparison with time-dependent methods is more straightforward. The operation count for LAD is approximately the same as the FTCS method using the same Poisson solver for the stream function Eq. (8.7.1b). Thus, the comparison of number of iterations for LAD and FTCS does realistically represent a computer-time comparison, and explicit time-dependent methods like FTCS may be categorically evaluated as inferior in terms of speed. If inefficient (for moderate to large problems) iterative methods like SOR or ADI [1] are used in the FTCS method for the inner solution of the Poisson equation, and the iterative convergence is adequate to produce even modest time accuracy, the evaluation is overwhelming. LAD can produce the final steady-state solution for large problems in less than the computer time for a *single time step* with FTCS.

8.14 Split NOS Method

A method that combines some advantages of both LAD and NOS is the Split NOS method [8-10]. The velocity vector V is split into an initial guess V^o and a perturbation V', not necessarily small. That is,

$$V^k = V^o + V'k. \qquad (8.14.1)$$

Then the steady-state form of the vorticity transport equation (8.7.1a) is solved for the k-th iteration again using (say) $0(\Delta x^2)$ centered differences, as

$$L(\zeta^k) = R_e \nabla \cdot (V' \zeta)^{k-1} \qquad (8.14.2a)$$

where the linear operator L is defined by

$$L(\zeta) = \nabla^2 \zeta - R_e \nabla \cdot (V^o \zeta) \qquad (8.14.2b)$$

Unlike the NOS method, this Split NOS method does not require an 'inversion' of the linear operator at each iteration, yet it retains *some* of the advection information of the NOS method. (The difference between LAD and Split NOS may be likened to the difference between analytical solutions obtained from low and high R_e expansions.) Surprisingly, the Split NOS method actually performs somewhat better, even in the iterations required, than the NOS method for the problem of developing channel flow.

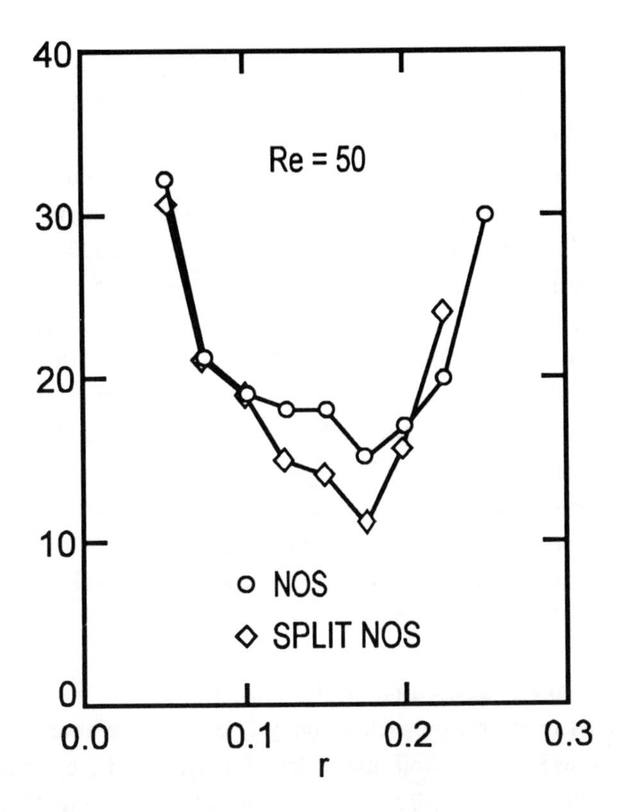

Figure 8.14.1. Comparison of convergence results between the NOS and Split NOS methods for the developing channel flow problem, $\Delta x = 1/10$, $\Delta y = 1$, $Re = 50$, Equation 8.9.1b for wall vorticity.

The initial guess used [8,9] for the velocity vector was just the inflow profile, i.e., $V^0 = (2y,0)$. The full range of variables previously described for LAD and NOS was used. A sample result is shown in Figure 8.14.1. At low R_e, the convergence values of NIT differed by no more than 1 between NOS and Split NOS. At high R_e, the Split NOS was better than NOS for r near the optimum; the best NIT for Split NOS was NIT = 11, compared to NIT = 15 for NOS. Similar results accrued for other R_e and other equations for wall vorticity [8,9].

The advantage of Split NOS over NOS is problem dependent, as may be shown by solutions of Burgers equation for the stagnation flow, $U(0) = 1, U(1) = 0, R_e = 10$. For a tolerance of $\epsilon = 10^{-5}$, NOS converges at NIT = 12, while Split NOS requires NIT = 17. For reversed boundary conditions (a suction-like flow with $U(0) = 0, U(1) = 1$) Split NOS converges at NIT = 13.

The Split NOS method compares to LAD and NOS as follows. For the flow-through problems of developing channel flow, the Split NOS method converges in slightly fewer iterations than NOS near optimum r, and in much fewer iterations than LAD. Like LAD, it does not require a new 'inversion' at each iteration, so that the relative number of iterations is a good indicator of relative computer times, for a single R_e case. However, different R_e problems cannot be stacked with Split NOS as they can with LAD, since a new influence coefficient matrix for L in equation (8.14.2) must be generated and inverted for each new R_e. Like NOS, the Split NOS method cannot use fast elliptic solvers limited to the Poisson equation, but for general V^0 requires less restricted solvers such as the marching method. However, for the example problem of developing channel flow with $V^0 = (2y,0)$, Fourier methods can be used with Split NOS; using the techniques clearly described by LeBail [12], the linear operator L of equation (10.3.2b) can be solved using fast Fourier transforms, provided that V^0 is of the form $V^0 = (f(y),\lambda)$ where λ is a constant. Unlike either NOS or LAD, the asymptotic convergence rate of Split NOS will depend somewhat on the adequacy of the initial guess for V^0; this disadvantage could be offset, at the cost of additional time to re-initiate the elliptic solver, by updating V° after a few iterations, or even periodically. (Of course, if V^0 is updated at each iteration, the Split NOS method reverts to NOS.) Also, note that V^0 containing reversed flow can be unstable [8,9]. The choice is not entirely clear cut, but the Split NOS method is generally preferable. (The LAD-like iterations will still be significant for other methods, including a 3-D method.)

8.15 A Better Boundary Condition on Wall Vorticity

The conventional forms for the wall vorticity ζ_w used above have a difficulty that can be overcome by the use of a method due to Israeli [17,18] and (separately) Dorodnicyn [19,20]. The conventional forms will still be of use with Semidirect methods that solve the linearized ζ and Ψ equations simultaneously (see Sections 8.18,19) but Israeli's method will give better iterative convergence for the 2-equation methods.

8.15.1 The Trouble with the Conventional Methods for Wall Vorticity

The computational boundary conditions used above for the vorticity at a no-slip wall present an anomaly. According to the analysis, the optimum relaxation factor r_{op} varies directly with the mesh spacing. So also does the maximum r for stable calculations, which for 1-D linear problems is just $2 \cdot r_{op}$. This means that as $\Delta Y \to 0$ then $r_{op} \to 0$, and the wall vorticity ζ_w would be held at its initial value ζ_x^o. The same behavior occurs for all three conventional forms for ζ_w mentioned in Section 8.8. For the linear 1-D problem, we still obtain the exact answer in one iteration for all finite Δy, but for non-trivial problems, iterative convergence is slower as Δx, Δy decrease.

The source of this anomaly of $r_{op} \propto \Delta y$ is not in the Semidirect method but in the conventional forms for wall vorticity. As pointed out to the present author by Thompson [21] and by Israeli [17,18], these finite-difference forms do not actually enforce the no-slip condition at the wall; although $u_w = v_w = 0$ is used in the derivation of these forms, a one-sided evaluation of u_w will not give $u_w = 0$ (except for trivial flows, or in the limit as $\Delta Y \to 0$).

In the present case, it is instructive to set-up the Semidirect solution for Poiseuille flow, as above, but using the continuum equations

$$\Psi_{yy} = \zeta, \quad \zeta_{yy} = 0 \tag{8.15.1}$$

instead of the finite-difference equations. In so doing, we find that the conventional boundary conditions on vorticity, like Eq. (8.9.1), simply reduce as $\Delta y \to 0$ to $\zeta = \Psi_{yy}$. This is *not* a boundary condition at all, but is simply the interior equation written on the boundary, which is no boundary condition at all. For finite Δy, the information that the wall is a no-slip surface (i.e., $u_w = v_w = 0$) is used in the conventional forms, but the contribution of this information drops out as $\Delta y \to 0$. In this limit, the Semidirect method gives a null iteration for ζ_w, which simply remains at its initial value. (Similar misgiving about these conventional equations for wall vorticity were expressed by Wu [22] and by Fix [23].)

8.15.2 Israeli-Dorodnicyn Method for Wall Vorticity

An alternate approach to ζ_w for finite-difference methods was given by Israeli [17,18], and by Dorodnicyn [19,20] in an early Semidirect calculation (see Section 8.16 below).

The idea is to evaluate u_w after each iteration by a one-sided finite-difference equation for $\delta\Psi/\delta n\,|_w$ and then iterate ζ_w until $\delta\Psi/\delta n\,|_w \simeq 0$. In Israeli's work, this was an intra-time-step iteration. In the present case, it is part of the overall Semidirect iteration. The conventional equations (8.9.1) for wall vorticity are replaced by

$$\zeta_w^k = g \cdot \frac{\delta\Psi}{\delta n}\Big|_w + \zeta_w^{k-1} \tag{8.15.2}$$

where $\delta\Psi/\delta n\,|_w$ is any consistent approximation to $\partial\Psi/\partial n\,|_w$. In [8-10] both the 2-point $0(\Delta y)$ approximation

$$\frac{\delta\Psi}{\delta n}\Big|_w = \frac{\Psi_{i,2} - \Psi_{i,1}}{\Delta n} \tag{8.15.3}$$

and a 3-point $0(\Delta y^2)$ approximation

$$\frac{\delta\Psi}{\delta y}\Big|_w = \frac{-\Psi_{i,3} + 4\Psi_{i,2} - 3\Psi_{i,1}}{2\Delta y} \tag{8.15.4}$$

were used. Note that g is not an under-relaxation parameter, in the sense that it does not weight the new value as some fraction of the old value.

8.15.3 Analytical Prediction of Optimum g

The optimum g can be easily predicted for simple flows based on *continuum* solutions rather than finite-difference solutions. For Poiseuille flow each solution of Eq. (8.12.1) gives the straight-line variation for $\zeta(y)$ as

$$\zeta(y) = (1-y)\,\zeta_w \tag{8.15.5}$$

where y = 1 at the centerline and $\Psi(1) = 1$, giving

$$\Psi(y) = (1-\frac{1}{3}\zeta_w)y + \frac{1}{2}\zeta_w y^2 - \frac{1}{6}\zeta_w y^3 \tag{8.15.6}$$

Since ζ_w is the only value that must converge, we evaluate u_w^{k-1} from Eq. (8.15.6) as

$$u_w^{k-1} = \frac{\partial \Psi}{\partial y}\Big|_{y=o}^{k-1} = 1 - \frac{1}{3}\zeta_w^{k-1} \tag{8.15.7}$$

(Note that the no-slip condition $u_w = o$ is met only when ζ_w converges to the correct value for Poiseuille flow, $\zeta_w = 3$.) The iteration Eq. (8.15.2) is then written as

$$\begin{aligned} \zeta_w^k &= g \cdot u_w^{k-1} + \zeta_w^{k-1} \\ &= g \cdot \left[1-\frac{1}{3}\zeta_w^{k-1}\right] + \zeta_w^{k-1} \\ &= g + (1-g/3)\,\zeta_w^{k-1} \end{aligned} \tag{8.15.8}$$

If g is taken as the correct steady-state value of 3, we get $\zeta_w^k = 3$ immediately, i.e., the optimum value is g = 3. With a more general normalizing scheme, we have

$$g_{opt} = 3\Psi_{max}/Y \tag{8.15.9}$$

In contrast to the conventional methods for wall vorticity that yield an iteration parameter r that depends strongly on Δy and that yield a null iteration as $\Delta y \to 0$, the Israeli-Dorodnicyn method yields an iteration parameter g that is not a function of Δy and an iteration scheme that behaves regularly as $\Delta y \to 0$.

It may be possible to use the inverted Aitken extrapolation procedure, as in [10], to estimate optimum g from early iterative behavior.

8.15.4 Performance of Israeli-Dorodnicyn Method

Using this method on the previous 2-D problem of developing channel flow, improved performance accrues even at coarse mesh spacing. The method converges in 6 iterations for $R_e \geq 100$ [10]. (This means that the correct solution is actually obtained at NIT = 5, and the calculation at NIT = 6 verifies iterative convergence.) Further, and most importantly, neither NIT nor g_{opt} are significantly affected by mesh refinements. This method was used in [24] to obtain accurate solutions of weakly separated flows using boundary-fitted coordinates, with up to 80×80 cells, including 4-the order accurate Richardson extrapolation.

8.16 Dorodnicyn-Meller Method

The Dorodnicyn-Meller [20] method may be regarded as a particular Split NOS method, with $\vec{V}^o = (u^o, v^o)$ limited to $u^o = u^o(y)$ and $v^o = 0$. They used the iteration scheme Eq. (8.15.8) for ζ_w, with optimum g based on the Poiseiulle flow analysis. Similar rapid convergence was obtained.

8.17 Viscous Flows in Alternate Variables

The Semidirect NOS, LAD, and Split NOS methods have been described economically in application to the Ψ-ζ system of variables, but are not limited to these. The principal ideas are applicable to primitive equations, velocity-vorticity variables [10] and to nonlinear problems other than fluid dynamics, i.e., glow-discharge lasers [25].

In the primitive u-v-P system of Navier-Stokes equations, a single operator L^o for Split NOS may be applicable to both the u and v momentum equations if boundary condition types are the same for both. The boundary conditions for the Poisson pressure equation at no-slip walls must be under-relaxed at low *Re* but the relaxation factor is not grid dependent. The Semidirect methods are applicable to both conservation and non-conservation forms, in collocated or staggered grids [10]. Iterative performance is comparable to the Ψ-ζ results [26]. Likewise for the velocity-vorticity system [10], which requires under-relaxation (grid-independent) of wall vorticity.

8.18 BID Method

The previous sections considered multidimensional Semidirect methods in which a scaler elliptic equation for each variable was solved sequentially, e.g. a linear equation for Ψ is solved, followed by a linearized equation for ζ. We now consider Semidirect methods in which the solution for the entire linearized *system* is obtained directly at each iteration, either by combining variables to a single higher-order equation, or by solving the coupled system directly. The first of these methods is the BID (BIharmonic Driver) method [9,27].

The results of the numerical experiments cited in Section 8.10 show the relative importance of the lagging boundary condition on vorticity. This introduces a *time-like* behavior into the iteration for the boundary values, and necessitates the determination of some optimum (or at least, stable) parameter. This is not difficult for simple flow-through problems like the developing channel flow, and performance is excellent for these problems with the Split NOS method. However, even with an optimum parameter, iteration is slower than anticipated for recirculating flow problems, represented by the driven cavity problem. This difficulty of the lagging boundary conditions is overcome in the BID method.

8.18.1 BID Iteration

Substituting the Poisson equation (8.7.1b) for stream function into the vorticity transport equation (8.7.1a), with $\nabla^2 \nabla^2 \Psi \equiv \nabla^4 \Psi$, gives

$$\frac{\partial}{\partial t}\left(\nabla^2 \Psi\right) = -R_e \nabla \cdot \left(V \nabla^2 \Psi\right) + \nabla^4 \Psi . \tag{8.18.1}$$

A method analogous to LAD immediately suggests itself. Setting the time derivative in Eq. (8.18.1) equal to zero, we define [9,27] the biharmonic driver method (abbreviated as BID) by

$$\nabla^4 \Psi^k = H^{k-1} = R_e \nabla \cdot \left(V^{k-1} \nabla^2 \Psi^{k-1}\right). \tag{8.18.2}$$

All the boundary conditions are incorporated into each BID iteration. For example, the conditions along a no-slip wall are

$$\Psi = 0, \quad \Psi_y = 0. \tag{8.18.3}$$

Thus, no under-relaxation of boundary values is required.

The second-order finite-difference operator used for the biharmonic operator is taken from [28], and is represented schematically in Fig. 8.18.1a for the case of $\Delta x = \Delta y = \Delta$.

Using centered differences for the right member of Eq. (8.18.2), the second-order finite-difference equation for interior points is

$$D^4 \Psi_{i,j}^k = \Delta^4 R_e \left\{ \frac{\frac{\delta \Psi}{\delta y}\Big|_{i+1,j} D^2 \Psi_{i+1,j} - \frac{\delta \Psi}{\delta y}\Big|_{i-1,j} D^2 \Psi_{i-1,j}}{\Delta x} - \frac{\frac{\delta \Psi}{\delta x}\Big|_{i,j+1} D^2 \Psi_{i,j+1} - \frac{\delta \Psi}{\delta y}\Big|_{i,j-1} D^2 \Psi_{i,j-1}}{\Delta y} \right\}^{k-1} \equiv G_{i,j}^{k-1} \tag{8.18.4a}$$

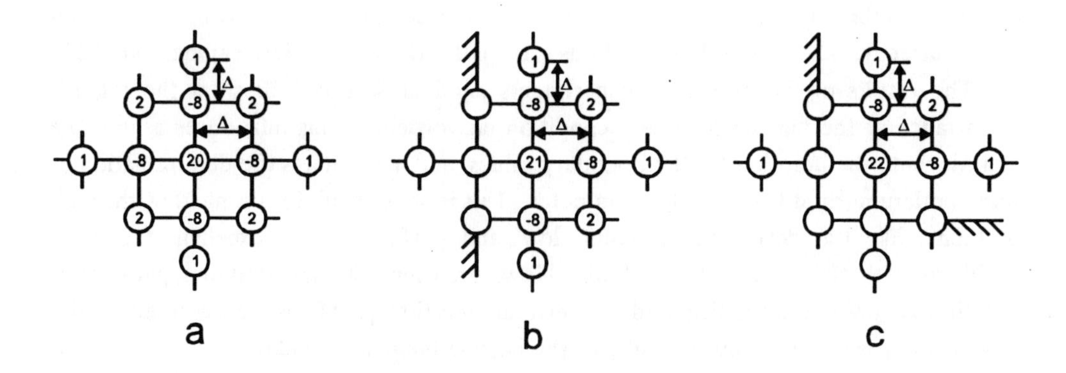

Figure 8.18.1. Finite difference biharmonic operators. (a) Operator for interior points. (b) Operator adjacent to left wall. (c) Operator adjacent to lower left corner.

where, for $\Delta x = \Delta y$,

$$
\begin{aligned}
D^4\Psi_{i,j} \equiv\ & \Psi_{i-2,j} + \Psi_{i,j+2} + \Psi_{i+2,j} + \Psi_{i,j-2} \\
& + 2\left[\Psi_{i-1,j+1} + \Psi_{i+1,j+1} + \Psi_{i+1,j-1} + \Psi_{i-1,j-1}\right] \\
& - 8\left[\Psi_{i-1,j} + \Psi_{i,j+1} + \Psi_{i+1,j} + \Psi_{i,j-1}\right] \\
& + 20\,\Psi_{i,j}
\end{aligned}
\tag{8.18.4b}
$$

$$
\left.\frac{\delta\Psi}{\delta y}\right|_{i\pm1} \equiv \frac{\Psi_{i\pm1,j+1} - \Psi_{i\pm1,j-1}}{2\Delta y} \tag{8.18.4c}
$$

$$
\left.\frac{\delta\Psi}{\delta x}\right|_{j\pm1} \equiv \frac{\Psi_{i+1,j\pm1} - \Psi_{i-1,j\pm1}}{2\Delta x} \tag{8.18.4d}
$$

$$
D^2\Psi_{\ell,m} \equiv \frac{\Psi_{i-1,j} + \Psi_{i,j+1} + \Psi_{i+1,j} + \Psi_{i,j-1} - 4\Psi_{i,j}}{\Delta^2} \tag{8.18.4e}
$$

When the above four equations with $\Delta x \equiv \Delta y$ are combined into Eq. (8.18.4a), the Δ terms all cancel.

8.18.2 BID Boundary Conditions for the Driven Cavity Problem

A great advantage of the BID method is that the steady-state boundary conditions are not lagged in the iterations. As a simple model problem representative of recirculating flows, we again consider the driven cavity problem shown previously in Fig. 8.9.1. The continuum boundary conditions are

$$
\Psi = 0 \quad , \quad \partial\Psi/\partial n = 0 \tag{8.18.5}
$$

along the sides and the bottom, and

$$
\Psi = 0 \quad , \quad \partial\Psi\,/\,\partial y = U \tag{8.18.6}
$$

along the lid. The finite-difference boundary conditions are set by aligning the walls at one node away from the center of the operator shown in Fig. 8.18.1a. For example, the left wall is positioned at i-1. The value of Ψ at i-1 is set as $\Psi_{i-1,j} = 0$. The value at i-2 is set by "reflection," with $\Psi_{i-2,j} = \Psi_{i,j}$, giving the consistent finite-difference gradient value of $\delta\Psi/\delta x\,|_{i-1} = 0$. These conditions are met by modifying the operator of

Fig. 8.18.1a to that of Fig. 8.18.1b for points adjacent to the left wall. Note that the weight of the center value has changed from 20 to 21. Near a corner, say the lower left corner, two such "reflections" are required, with $\Psi_{i-2,j} = \Psi_{i,j-2} = \Psi_{i,j}$. Also, both $\Psi_{i-1,j} = 0$ and $\Psi_{i,j-1} = 0$. Consequently, the operator of Fig. 8.18.1a is modified to that of Fig. 8.18.1c at the point adjacent to the lower left corner; note that the weight of the center value is now 22 instead of 20.

The velocity of the moving lid is U. Near the lid, we set the consistent finite difference gradient value as

$$\delta\Psi/\delta y\,|_{i,j+1} \equiv \frac{\Psi_{i,j+2} - \Psi_{i,j}}{2\Delta y} = U \tag{8.18.7}$$

or

$$\Psi_{i,j+2} = \Psi_{i,j} + 2\Delta y \cdot U \tag{8.18.8}$$

To accommodate the lid velocity $U \neq 0$, it is only necessary to modify the biharmonic operator as in Fig. 8.18.1b (rotated), or as in Fig. 8.18.1c (rotated) near the corners, and to add the velocity term of Eq. (8.18.8) into the nonhomogeneous right member of Eq. (8.18.4a), which then becomes

$$D_L^4 \Psi_{i,j}^k = G_{i,j}^{k-1} - 2\Delta y \cdot U \tag{8.18.9}$$

where D_L^4 is the operator of Fig. 8.18.1b or -c, properly rotated.

It is possible to use a marching method for the biharmonic equation (see Chapter 2, Sections 2.4 and 2.5) but two other direct biharmonic solvers were used in [27]; these are due to Bauer and Reiss [29] and Buzbee and Dorr [30].

8.18.3 BID Performance for the Driven Cavity

To compare the performance of BID with the NOS and LAD methods, we quantify the iterative convergence criteria as in Section 8.9, requiring

$$\zeta^k - \zeta^{k-1} < \epsilon = 10^{-5} \tag{8.18.10}$$

where the summation extends only over the interior points. The vorticity is evaluated from the Poisson equation (8.7.1b).

The problem sizes studied in [27] varied from 8×8 cells to 100×100 cells.Starting with initial conditions of $\Psi = 0$ everywhere, the first iteration for any R_e gives the Stokes flow solution, corresponding to $R_e = 0$. At $R_e = 10$, the solution is attained at NIT = 5 and verified by the criterion of Eq. (8.18.10) at NIT = 6. At $R_e = 20$, convergence is verified at NIT = 8. This convergence rate is not affected by mesh size for $\Delta = 1/10$ to $1/100$, which is a considerable advantage. The results are presented in Table 8.18.1. The deterioration at $R_e = 50$ is possibly due to cell R_e effects, or possibly just the iteration scheme deteriorating as does LAD.

Re	K	Δ
0	2 (1)	1/10 - 1/100
10	6	1/10 - 1/100
20	8	1/10 - 1/100
50	14-18	1/10 - 1/100

Table 8.18.1. Convergence rate for BID on the Driven Cavity problem.

This iteration convergence rate is superior to the NOS and LAD methods for this recirculating flow problem and is, of course, far superior to any explicit time-dependent or time-like method. Also of great advantage is the fact that no boundary under-relaxation factor need be ascertained, as it must be in the NOS and LAD methods, nor any time step, as in the time-dependent methods.

It is noteworthy, however, that blind application of iterative convergence criteria like Eqn. 8.18.10 can be deceptive. At Re = 20, the inequality (8.18.10) was satisfied at NIT = 2, but not at NIT = 3, 4, ... until "true" convergence was verified at NIT = 8. In [27], we examined the behavior beyond the values of NIT stated for convergence. As a practical recommendation, we found that requiring NIT > 2 in addition to Eqn. 8.18.10 was sufficient to ensure "true" convergence.

Both the Bauer-Reiss [29] and the Buzbee-Dorr [30] biharmonic solvers give answers to essentially the single precision of the computer, ~14 significant figures in [27]. The computer times are roughly comparable to the times for a single time step using an explicit time-like method with an analogous fast Poisson solver. The Bauer-Reiss method has a higher initialization time but a lower computer time for subsequent solutions. Problems with different R_e do not require re-initialization of the biharmonic solver; different R_e cases can be stacked (in a continuation method) in the same run without separate initializations, provided that the grid size and the class of boundary conditions are not changed. For example, the driven cavity problem might be run with any number of moving walls or moving segments of walls without re-initialization, since the wall values of $\Psi = 0$ are always specified and the different values of wall speeds can be accommodated into the non-homogeneous terms as in Eq. (8.18.9). Likewise, problems with symmetry boundaries or outflow boundaries will not require new initializations of the biharmonic solver. For large stacks of cases, the Bauer-Reiss method will be faster, especially at large Δx. However, its storage requirements are more severe. (For a large $M \times M$ cell problem, the Bauer-Reiss method requires approximately $3M^3$ storage locations, while the Buzbee-Dorr method requires only about $3M^2$.)

The vorticity along the driving lid was evaluated in [27] using the $0(\Delta^2)$ Woods' equation (8.9.1b). This vorticity does *not* enter in to the dynamic iteration process, but is merely calculated in post-processing. Accuracy comparisons are given in [27] . The BID solutions are second-order accurate, as are the Ψ-ζ solutions, but the BID solutions are not equal to the Ψ-ζ solutions for finite Δ. This difference is due to the different continuum equations used. For example, it may be verified by hand calculations that the Ψ-ζ equations, using a second-order form such as Eqn. (8.9.1b) for the wall vorticity, give the exact continuum solution for Poiseuille flow but the biharmonic solution does not. Comparing the BID and Ψ-ζ solutions for the driven cavity [27], it appears that the Ψ-ζ approach is more accurate for this two-dimensional problem also.

The computer time required to obtain a solution with BID obviously depends on the efficiency of the biharmonic solver itself. It is also perfectly obvious that the number of iterations should not depend on the biharmonic solver. Mittal and Sharma [31] have confused the literature on this point. They used an iterative method (using "the idea of pre-conditioned conjugate gradient") as the linear biharmonic solver, and claimed that the number of iterations required to reach convergence was reduced from that reported in [27]. It would be remarkable (though not strictly impossible) if the incomplete iterative convergence of an iterative method for the linearized equations speeded the nonlinear iteration. A more likely candidate is a coding error. (The first draft of [31] indicated even better performance than the final published version, but examination of the code listing disclosed an error; the authors had failed to re-set the initial conditions in a stack of R_e runs, so that each successive case was starting from a very good initial condition, yet was being compared to [27] in which each run started from initial conditions of zero flow. The authors of [31] re-calculated their performance runs with apparently yet another error in the code, and published the results.) It is an equally obvious point that if an iterative solver is to be used, it makes more sense to *not* converge it well for each linear solution but rather to frequently update the linearization within the iteration.

An excellent implementation of a fast direct solver coupled iteratively via Newton and chord iterations is given by Lippke and Wagner [32], who also use high-accuracy stencils for the convection terms, appropriate to achieve a "balanced method" [33] at high R_e.

8.18.4 BID Performance for Flow-Through Problems

The BID method may also be adapted to flow through problems [Roache and Ellis, 27]. The inflow and symmetry boundaries involve straightforward stencil modifications given in detail in [27]. The outflow boundary is treated analogously to the out-flow condition in the Ψ-ζ system. Assuming $(\partial^2/\partial x^2)\,(\;) = 0$ gives the fourth-order ordinary differential equation at out-flow

$$\frac{d^4\Psi}{dy^4} = G \tag{8.18.11}$$

where G is defined in Eq. (8.18.4a). The outflow boundary is set at $i = IL-1$, and $\partial G/\partial x = 0$ is approximated by setting $G_{IL-1} + G_{IL-2}$. Then five-point$O(\Delta y^2)$ differences for Eq. (8.18.11) gives a pentadiagonal matrix that is solved directly by a general bounded matrix solver [34,35] or by a special pentadiagonal solver analogous to tridiagonal solvers. Other details of the boundary conditions can be found in [27].

Like the LAD method, the BID method has no advection information in the linear operator; like LAD, it fails to converge at higher R_e for flow through problems. The developing channel flow problem was solved in only 4 iterations at $R_e = 1$ and $R_e = 10$, but did not converge at $R_e = 20$.

8.19 FOD and Coupled Systems Solvers

Methods analogous to the NOS and Split NOS methods in Ψ-ζ immediately suggest themselves. The BID is analogous to the LAD method; a method analogous to NOS would be the Fourth-Order Driver (FOD) method.

$$\nabla^4 \Psi^{k+1} - R_e \nabla \cdot \left(V^k \nabla^2 \Psi^{k+1}\right) = 0 \tag{8.19.1}$$

Similarly, a method analogous to Split NOS would be the Split FOD method,

$$\nabla^4 \Psi^{k+1} - R_e \nabla \cdot \left(V^0 \nabla^2 \Psi^{k+1}\right) = R_e \nabla \cdot \left(V'^k \nabla^2 \Psi^k\right) \tag{8.19.2}$$

wherein total velocity $V = V^0 + V'$, and V^0 is an initial guess. By analogy with the results for the LAD and NOS methods, we infer that these methods would undoubtedly accomplish the high R_e solutions for flow-through problems. For the FOD method, each iteration would require a new initialization of the linear solver, so that this method would certainly be more time-consuming than BID. However, the Split FOD method would not require initialization at each iteration. (Like the Split NOS method, it would require re-initialization for each R_e case.)

Each of these would require the direct linear solution of fourth-order elliptic equations with variable coefficient second derivatives. The method of the Buzbee-Dorr [30] solver is not so general, but the method of the Bauer-Reiss [29] solver can be adapted, with modifications, to a general block-5 matrix solver. Also, the marching methods can be adapted to higher-order equations as in Chapter 2, Sections 2.4, 2.5. However, the marching methods are better adapted to simultaneously solving the linearly coupled second-order equations for Ψ and ζ, rather than a single fourth-order equation for Ψ. The operation count is less, the truncation error is improved, the boundary conditions are easier to implement, and the method is more readily extended to higher dimensions. In either case, the essential advantage of treating the fourth-order system is retained: that of requiring no nested iteration for the wall vorticity.

One final remark in regard to fourth-order systems is appropriate in the light of recent work on "mixed" methods, which have been motivated by the observation that velocities may converge (in truncation error, not iteration error) slower than other variables. For example, it is possible to have a consistent discretization system based on finite elements in which pressure or stream function convergence is second order in space, but velocity convergence is only first order in space. This can occur even when the more forgiving L^2 norm is used to measure the discretization error. But it is shown in [27] that for BID (and therefore for other fourth-order systems such as the FOD method) convergence is second order in space for stream function, velocities, and vorticity, even in the more difficult L^∞ norm.

8.20 Other Applications and Non-Time-Like Methods

In addition to the fluid dynamics applications cited earlier, Semidirect methods have been applied to quasi-steady flow of double-diffusive bouyancy problems in salt-gradient solar ponds [36], in nonlinear electric field calculations for lasers [25,37], for boundary-fitted non-orthogonal grid generation [38].

Other methods that avoid the time-like iteration for boundary values are mentioned below. As anticipated, they have excellent convergence properties.

Morihara and Cheng [39] used a rapidly-converging Semidirect method to calculate developing channel flow using a form of the Navier-Stokes equations in which pressure has been eliminated and the velocities appear up to third-order derivatives. The iteration scheme was like the NOS method with the added feature that convergence of the nonlinear terms is speeded by quasilinearization. Like the NOS method, their method required the *re-initialization* of a Gaussian elimination routine at each iteration. They actually use Gaussian elimination to solve the 2-D linear algebra problem directly, rather than a marching solution, so the solutions are expensive. But even if marching methods were used, the quasilinearization would require re-initialization at each iteration, like the NOS method.

Based on the experiments cited in Section 8.10 on the importance of lagging boundary values, it appears that a real advantage of the Morihara-Cheng method is in the use of the velocity variables, which have *known* steady-state boundary conditions. Thus, no time-like iteration occurs at the boundaries.

An obvious candidate method is the combination of the Morihara-Cheng method with the basic linearization idea of the Split NOS method. This method would use velocity variables, giving known steady-state boundary conditions to speed iteration convergence, yet the linear algebra routine would be initiated only once for each R_e case, thus offsetting the increase in number of iterations by a decrease in computer time.

This chapter's primary focus has been on Semidirect methods that can utilize spatial marching methods for elliptic equations. The elliptic Marching Methods cannot be applied to (linearized) versions of the Euler equations, nor the primitive Navier-Stokes compressible equations with the usual hyperbolic form of the continuity equation, since these are not second-order equations. There are other non-time-like nonlinear iterative methods that are likewise exceptionally efficient.

The Davis methods for and interacting boundary layer solutions [40] and the full Navier-Stokes equations [41] are extremely efficient Semidirect methods that utilize nonlinear direct solutions (e.g., classical boundary layer) as the base solution, and again lag corrections for the full (Navier-Stokes) terms. (Davis's interacting boundary layer solutions [40] were the original inspiration for the present author's Semidirect methods [6].) Martin and Lomax [42] and Martin [43] used Semidirect methods (and coined the term) for subsonic and transonic aerodynamics flows.

The computational fluid dynamics literature contains many steady-state methods that use the familiar full Newton-Raphson nonlinear iteration. Most of these are finite element solutions [e.g., 44-45] but there are as well finite difference methods [46-48]. Although these converge very rapidly in terms of number of iterations once the asymptotic regime is reached, we generally prefer the presently described type of Semidirect methods because of their greater efficiency as measured by computer time, especially for moderate iterative convergence tolerances.

The "Semidirect" approach can also be used for linear problems, using iterations to lag corrections to terms not treatable by some chosen direct linear solver, e.g., see Ehrlich [14]. (Note that this philosophy extends to *anything* that you know how to solve directly, and thus the "Semidirect" family overlaps ADI [1], in which the direct solver is just the tridiagonal algorithm.) For linear problems, the Semidirect method would usually be described in terms of "preconditioning."

8.21 Remarks on Solution Uniqueness

The preceding discussion of *linear* Semidirect problems provides a reminder about nonlinearity. In the Introduction to this Chapter, we noted the obvious limitation of the underlying assumption of Semidirect methods, that a steady-state solution exists. It is appropriate to note here the possible problem of non-uniqueness, i.e. the existence of *more* than one steady-state solution. Multiple *physical* solutions certainly exist in fluid dynamics (e.g. airfoil stall hysterisis, as discussed in [1, p. 6-7]) and certainly in other disciplines involving nonlinearities. The question posed in [1] is still pertinent and unanswered: Towards *which* solution, if any, would a numerical scheme converge?

In addition to physical non-uniqueness, the discretization of nonlinear problems can intoduce more severe non-uniqueness. This is shown by the following trivial example.

Consider the nonlinear Burgers Equation (8.4.1) discretized with the minimum possible grid of $2\Delta x$. We use upstream differencing of the advection term, which should give the best possible qualitative behavior for this minimum discretization. This gives

$$(u_3 - 2\,u_2 + u_1\,)\;/\;\Delta x^2 - R_e\,u_2\,(u_2 - u_1)\;/\;\Delta x \;=\; 0 \tag{8.21.1}$$

With $\Delta x = 1/2$ and boundary conditions of $u_1 = 0$ and $u_3 = 1$ this gives

$$(R_e\;/\;2)\;u_2^2 + 2\,u_2 - 1 \;=\; 0 \tag{8.21.2}$$

$$u_2 = \{-2 \pm 2\sqrt{(1 + R_e/2)}\;\}/\;R_e \tag{8.21.3}$$

For $Re = 1$, the two solutions give the reasonable $u_2 \approx +\,0.45$ and the ridiculous $u_2 \approx -\,4.45$.

For finer discretization, the multiplicity of discrete solutions increases, and generally we could expect n (possibly non-distinct) solutions where n is the number of interior nodes. Fortunately, the practical situation is not so daunting. In a most significant paper, Shubin et al. [49] have demonstrated that the multiple solutions for a blunt body shock problem converge to a single solution as the mesh size approaches 0. Nevertheless, we cannot entirely ignore the possibility of non-physical non-uniqueness.

Note, however, that this problem is *not* limited only to Semidirect nonlinear methods, since the same possibilities for physical and non-physical non-uniqueness exist for transient and pseudo-transient methods.

8.22 Remarks on Semidirect Methods within Domain Decomposition

Finally, we note that Semidirect methods utilizing marching methods for the linearized problems are naturally suited to Domain Decomposition based on geometry and/or equation types, e.g., viscous and inviscid regions, as in Dinh et. al. [50] and earlier Wu [51]. (See also [52] and the references and discussion in Chapter 3.) Since the nonlinear problem is essentially iterative even without Domain Decomposition, iterative coupling of subdomains is a natural choice. (See Chapter 3, Section 3.8 and Chapter 5, Section 5.12.)

It is significant that geometric Domain Decomposition methods using Picard-like iterations work virtually as well on mild nonlinearities (such as the Navier-Stokes equations without shocks) as on linear problems. The *analysis* of iterative convergence is greatly affected, but the performance is not. Parallelization is often natural, as discussed in Chapters 3 and 7. Thus the Semidirect methods presented in this chapter are good candidates for parallel solutions of nonlinear problems via Domain Decomposition.

References for Chapter 8

1. P. J. Roache, *Computational Fluid Dynamics*, rev. printing, Hermosa Publishers, Albuquerque, NM, 1976.
2. H. F. Weinberger, *Partial Differential Equations,* Blaisdell Publishing Co., New York, 1965, p. 51.
3. R. W. Hockney, A Fast Direct Solution of Poisson's Equation Using Fourier Analysis, *Journal of the Association for Computing Machinery,* Vol. 12, 1965, pp. 95-113.
4. P. J. Roache, A Direct Method for the Discretized Poisson Equation, Sandia National Laboratories, Report SC-RR-70-579, Albuquerque, NM, 1971.
5. P. J. Roache, A New Direct Method for the Discretized Poisson Equation, in M. Holt, (ed.), *Lecture Notes in Physics, Vol. 8, Proc. Second International Conference on Numerical Methods in Fluid Mechanics*, Springer, New York, 1971, pp. 48-53.
6. P. J. Roache, Finite Difference Methods For The Steady-State Navier-Stokes Equations, Sandia National Laboratories, Report SC-RR-72-0419, Albuquerque, NM, Dec. 1972.
7. P. J. Roache, Finite Difference Methods for the Steady-State Navier-Stokes Equations, *Lecture Notes in Physics*, Vol. 18, Springer, New York, 1973, pp. 138-145.
8. P. J. Roache, The LAD, NOS and Split NOS Methods for the Steady-State Navier-Stokes Equations, *Computers and Fluids*, Vol. 3, 1975, pp. 179-195.
9. P. J. Roache, The Split NOS and BID Methods for the Steady-State Navier-Stokes Equations, In R. D. Richtmyer, (ed.), *Lecture Notes in Physics, Vol. 35, Proc. Fourth International Conference on Numerical Methods in Fluid Dynamics*, Springer, New York, 1975, pp. 347-352.
10. P. J. Roache, A Semidirect Method for Internal Flows in Flush Inlets, AIAA Paper 77-647, *Proc. AIAA Third Computational Fluid Dynamics Conference*, 1977.

11. H. Schlichting, *Boundary-Layer Theory*, translation J. Kestin, McGraw-Hill , New York, Sixth Edition, 1968, pp. 107-108.
12. O. Buneman, A Compact Non-Iterative Poisson Solver, SUIPR Rept. 294, Stanford University, Stanford, CA, May 1969.
13. W. R. Briley, *A Numerical Study of Laminar Separation Bubble using the Navier-Stokes Equations*, Report J110614-1, United Aircraft Research Laboratories, Hartford, CN, 1970.
14. L. W. Ehrlich, Solving the Biharmonic Equation as Coupled Finite Difference *Equations, SIAM Journal of Numerical Analysis*, Vol. 8, No. 2, June 1971, pp. 278-287.
15. J. H. Ferziger, Estimation and Reduction of Numerical Error, *FED Vol. 158, Symposium on Quantification of Uncertainty in Computational Fluid Dynamics,* ASME Fluids Engineering Division, Summer Meeting, Washington, D.C., June 20-24, 1993. I. Celik, C. J. Chen, P. J. Roache, and G. Scheurer, (eds.), pp. 1-7.
16. D. C. Thoman and A. A. Szewczyk, Time Dependent Viscous Flow Over a Circular Cylinder, *The Physics of Fluids Supplement II*, 1969, pp. 76-87. See also D. C. Thoman and A. A. Szewczyk, *Numerical Solutions of Time Dependent Two Dimensional Flow of a Viscous, Incompressible Fluid Over Stationary andRotating Cylinders*, Technical Report 66-14, Heat Transfer and Fluid Mechanics Laboratories, Dept. of Mechancial Engineering, University of Notre Dame, Notre Dame, IN, July 1966.
17. M. Israeli, *Studies in Applied Mathematics*, Vol. 49, 1970, p. 327. See also M. Israeli, *Studies in Applied Mathematics*, Vol. 51, 1972, p. 67.
18. S. A. Orszag and M. Israeli, Numerical Simulation of Viscous Incompressible Flows, *Annual Review of Fluid Mechanics*, Vol. 6, 1974, pp. 281-318.
19. A. A. Dorodnicyn, *A Review of Methods for Solving the Navier-Stokes Equations, Third International Conference on Numerical Methods in Fluid Mechanics*, Springer, New York, 1973, pp. 1-11.
20. A. A. Dorodnicyn and N. A. Meller, Application of the Small Parameter Method to the Solution of Navier-Stokes Equations, *Fluid Dynamics Transactions*, Vol. 5, Part II, 1971, pp. 67-82.
21. J. F. Thompson, personal communication, 1976.
22. J. C. Wu, Numerical Boundary Conditions for Viscous Flow Problems, *AIAA Journal*, Vol. 14, No. 8, 1976, pp. 1042-1049.
23. G. J. Fix, Numerical Solution of Fourth Order Equations Arising in Incompressible Flow Problems, *Proc. Fifth International Conference on Finite Elements in Water Resources*, University of Vermont, Burlington, VT, June-18-22, 1984, J. P. Laible, C. A. Brebbia, W. Gray and G. Pinder, (eds.), Springer-Verlag, New York, 1984, pp. 807-813.
24. P. J. Roache, Scaling of High-Reynolds-Number Weakly Separated Channel Flows, *Numerical and Physical Aspects of Aerodynamic Flows*, T. Cabeci, (ed.), Springer-Verlag, New York, 1982, pp. 87-98.
25. P. J. Roache, S. Steinberg, and W. M. Moeny, Interactive Electric Field Calculations for Lasers, AIAA Paper 84-1655, *AIAA 17th Fluid Dynamics, Plasma Physics, and Lasers Conference*, 25-27 June 1984, Snowmass, CO.

26. P. J. Roache, 1978, unpublished.
27. P. J. Roache and M. A. Ellis, The BID Method for the Steady-State Navier-Stokes Equations, *Computers and Fluids*, Vol. 3, 1975, pp. 305-320.
28. M. G. Salvadori and M. L. Baron, *Numerical Methods in Engineering*, Prentice-Hall, Englewood Cliffs, NJ, Second Edition, 1961.
29. L. Bauer and E. L. Reiss, Block Five Diagonal Matrices and the Fast Numerical Solution of the Biharmonic Equation, *Mathematics of Computation*, Vol. 26, No. 118, 1972, pp. 311-326.
30. B. L. Buzbee and F. W. Dorr, *The Direct Solution of the Biharmonic Equation on Rectangular and the Poisson Equation on Irregular Regions*, LA-UR-73-636, Los Alamos Scientific Laboratories, NM, 1973. See also SIAM *Journal of Numerical Analysis*, Vol. 11, 1974, pp. 753-763.
31. R. C. Mittal and P. K. Sharma, Fast Finite Difference Solution for Steady-State Navier-Stokes Equations Using the BID Method, *International Journal for Numerical Methods in Fluids*, Vol. 7, 1987, pp. 911-917.
32. A. Lippke and H. Wagner, A Reliable Solver for Nonlinear Biharmonic Equations, *Computers and Fluids*, Vol. 18, No. 4, 1990, pp. 405-420.
33. P. J. Roache, A Pseudo-Spectral FFT Technique for Non-Periodic Problems, *Journal of Computational Physics*, Vol. 27, No. 2, May 1978, pp. 204-220.
34. J. J. Dongarra, C. B. Moler, J. R. Bunch, and G. W. Stewart, *LINPACK User's Guide*, Society for Industrial and Applied Mathematics, Philadelphia, 1979.
35. E. Anderson et al., *LAPACK Users' Guide*, Society for Industrial and Applied Mathematics, Philadelphia, 1992.
36. P. J. Roache, A Preliminary Study of Double Diffusion Driven Natural Convection in a Salt Gradient Solar Pond, unpublished, 1978.
37. M. Von Dadelszen, W. M. Moeny, and P. J. Roache, Electric Field Calculations Using the ELF Codes, *Proc. IEEE Pulsed Power Conference*, June 10-12, 1985, Crystal City, DC.
38. P.J. Roache, Semidirect/Marching Methods and Elliptic Grid Generation, *Proc. Symposium on the Numerical Generation of Curvilinear Coordinate Systems and Use in the Numerical Solution of Partial Differential Equations*, April 1982, Nashville, TN, J. F. Thompson, ed., North-Holland Publishing Co., Amsterdam, pp. 729-737.
39. H. Morihara and R. T. Cheng, Numerical Solution of the Viscous Flow in the Entrance Region of Parallel Plates, *Journal of Computational Physics*, Vol. 11, 1973, pp. 550-572.
40. R. T. Davis and M. J. Werle, Numerical Solutions for Laminar Incompressible Flow past a Paraboloid of Revolution, *AIAA Journal*, Vol. 10, 1972, pp. 1224-1230.
41. R. T. Davis, Numerical Solutions of the Navier-Stokes Equations for Laminar Incompressible Flow past a Parabola, *Journal of Fluid Mechanics*, Vol. 51, 1972, pp. 417-433.
42. E. D. Martin and H. Lomax, Rapid Finite-Difference Computation of Transonic Aerodynamic Flows, AIAA Paper No. 74-11, 1974.

43. E. D. Martin, A Generalized Capacity-Matrix Technique for Computing Aerodynamic Flows, *Computers and Fluids*, Vol. 2, 1974, pp. 79-97.
44. D. K. Gartling and P. J. Roache, Efficiency Trade-Offs on Steady-State Methods Using FEM and FDM, *Proceedings First International Conference Numerical Methods in Laminar and Turbulent Flow*, 17-21 July 1978, University College, Swansea, Wales, Pentech Press, London, pp. 103-112.
45. O. C. Zienkiewicz and R. L. Taylor, *The Finite Element Method*, Vol. 1, 1989, Vol. 2 1991, 4th Edition, McGraw-Hill, New York.
46. B. Fornberg, A Numerical Study of Steady Viscous Flow Past a Circular Cylinder, *Journal of Fluid Mechanics*, Vol. 98, part 4, 1980, pp. 819-855.
47. R. Schreiber and H. B. Keller, Driven Cavity Flows by Efficient Numerical Techniques, *Journal of Computational Physics*, Vol. 49, 1983, pp. 310-333.
48. K. T. Walter and P. S. Larsen, The FON Method for the Steady Two-Dimensional Navier-Stokes Equations, *Computers and Fluids*, Vol. 9, 1981, pp. 365-376.
49. G. R. Shubin, A. B. Stephens, H. M. Glaz, A. B. Wardlaw, and L. B. Hackerman, Steady Shock Tracking, Newton's Method, and the Supersonic Blunt Body Problem, *SIAM Journal of Scientific and Statistical Computing*, Vol. 3, No. 2, June 1982, p. 127.
50. Q. V. Dinh, R. Glowinski, J. Periaux, and G. Terrason, On the Coupling of Viscous and Inviscid Models for Incompressible Fluid Flows via Domain Decomposition, in R. Glowinski et al. (Ref. 52), 1988, pp. 350-369.
51. J. C. Wu, Separate Treatment of Attached and Detached Flow Problems, *AIAA Journal*, Vol. 19, No. 1, 1981, pp. 20-27.
52. R. Glowinski, G. Golub, G. A. Meurant, and J. Periaux, *First International Symposium on Domain Decomposition Methods for Partial Differential Equations*, Ecole Nationale des ponts et Chaussees, Paris, Jan, 1987; SIAM, Philadelphia, 1988.

Chapter 9

COMPARISON TO MULTIGRID METHODS

9.1 Introduction

A. Brandt [1] has pioneered the development of a class of methods, called "multigrid methods", which have been very successful in a wide range of applications. Although iterative rather than direct, under ideal conditions they can approach the speed of direct methods, depending on the iterative convergence criterion used. The multigrid literature is now huge, e.g., see [1-13], and it is appropriate here to consider, if only primarily in a qualitative way, the comparison of marching methods to multigrid methods [14]. Although both methods share the common attribute of speed, they differ significantly in many other points of comparison. For clarity, we have arranged the discussion around each point of comparison, although there are overlaps and repetitions. As a convenient shorthand notation in this chapter only, we will abbreviate (elliptic) "marching method(s)" by MM and "multigrid method(s)" by MG. By "basic MM," we refer to the unstabilized marching method (Chapters 1,2,5) without iterative corrections, patching, (Chapter 3), etc.

9.2 Definition of the Methods

The terms "algorithm," "method," "families of methods," "methodology," etc. are often used in different senses, and are often confused with "codes." Without attempting to be unnecessarily precise, we can say that MM, although they certainly represent a family of methods and that more than one algorithm and/or code are possible for each specific method, tend to be more specifically defined than MG. (One MG researcher describes multigrid as a philosophy, rather than a method.) This is not at all a criticism of MG, and in fact accounts partially for its extraordinary fruitful applications. But it also accounts for the fact that competent workers have failed on particular problems that later proved to be well suited to MG, and that others have spent literally *years* of development time to achieve that potential performance. This is notably the case in transonic flow calculations, Euler solvers, and two-phase flow in porous media [15].

9.3 Treatment of Nonlinearities

MM, per se, do not treat nonlinear problems. These can be treated by a combination of MM and some kind of semidirect nonlinear iteration; I generally prefer Picard and quasi-Picard nonlinear iterations (Chapter 8). MG can be used in this mode, as an inner linear solver for various outer nonlinear iteration methods, but it is more efficiently used as a nonlinear iteration method; see the FAS (for Full Approximation Scheme) Cycle C algorithm in [1]. (Although the FAS scheme is more difficult to understand than the Linear Cycle C, we paradoxically have found it easier to debug.) It is difficult to set up a "black box" MG [10] to make use of this feature on general user-defined nonlinear problems, and at the time of this writing, it has not been accomplished. Even for specific problems, researchers have often chosen to use MG only as an inner linear solver rather than attempt a full nonlinear MG. MM is not applicable to a Newton-Raphson procedure; MG is, but is difficult and can be fragile.

9.4 Speed and Accuracy

"Accuracy" in the present context refers to accuracy of the solution of the discretized equation, as measured perhaps by residuals, rather than to accuracy of the discretization, as measured perhaps by truncation error.

The speed of the basic MM, like that of any direct method, is fixed by the grid and the equation being solved. The accuracy that results from a calculation does not depend on the speed or vice versa; it is fixed for a given computer. (However, this is modified somewhat for iterative corrections, which are recommended (Chapter 6, Section 6.8) only for the 9-point problem with periodic boundary conditions, and for subgrid solutions by MM, and Domain Decomposition using iterative coupling of MM solutions, which are related to MG.) For MG, like any iterative method, there is a relation between speed and accuracy, although it is the great attraction of MG that the two are not so strongly coupled as is the case with simple iterative methods. For the prototype Poisson problem on a rectangular grid, MG iteration reduces the error (from the initial guess or from the previous cycle) by about an order of magnitude for every 4-10 iterations. It is claimed that MG operation count is optimal, i.e. merely proportional to the number of unknowns, but this depends on the iteration being terminated at a level set by estimates of the truncation error. If tighter convergence criteria are used, the methods will not be optimal. However, even with a tight convergence criterion, the number of iterations required does not increase nearly as rapidly with mesh refinement as SOR or even ADI [16].

For linear problems, the basic MM will be faster for repeat solutions. For initial solutions and for nonlinear solutions, MG will be as fast or faster, for strongly elliptic problems, depending on the success of finding a good "smoothing" algorithm for the equations and grid; experience indicates that a good smoother can usually be found, although it may be difficult; see Section 9.2 above. For mixed, weakly elliptic or non-elliptic problems, MG success is dependent on finding good restriction and prolongation operators (interpolating in the most general sense). This can be very difficult.

A recurring claim in some modern papers on iterative methods, including but not limited to MG, e.g., see [1,3], is that iterative convergence should not be pursued beyond the level of truncation error. That is, it makes no sense to spend the computer time to reduce residuals to (say) machine single-precision zero (i.e., to the point of incipient or actual underflow) unless the computational (fine) grid is fine enough to provide truncation error levels comparable to machine precision. Or so the argument goes.

There certainly is a correlation, e.g., if an iterative tolerance of $\epsilon = 10^{-4}$ is necessary in a 20×20 grid, then something much smaller will be necessary in a 100×100 grid. However, building codes to implement this relation, and basing operation counts on same, has some serious difficulties.

First and most obvious to practice-oriented, real problem-solving users, is that the quantative relation between the size of the residual and the size of the truncation error is not evident. Even if a residual norm is selected (a much more delicate choice than most theoreticians admit) and if local truncation errors are estimatable (not at all clear for first derivative terms present, aggravated by non-orthogonal grids and large flow angles with respect to grid lines), one still has the formidable problem of establishing global truncation errors, which are certainly problem dependent.

A less obvious point is the contamination of grid convergence testing, especially if Richardson extrapolation is used (e.g., see Roache [16-18], Roache and Knupp [19]). Incomplete iteration errors (i.e., residual errors) are magnified in the extrapolation, which can invalidate truncation error estimates (imagine the deleterious feedback possible in solution-adaptive grid generation) or even produce an extrapolated (to $\Delta X \rightarrow 0$) solution that is *less* accurate than the fine grid solution obtained by iteration.

As another example, consider the following problem encountered by my colleagues [20] working on the simulation of transport of radioactive contaminants by flow in porous media. The flow and transport problems are decoupled. For the flow problems, residuals were reduced to $0(10^{-7})$ in a roughly 50 $\times$ 80 cell grid, but the velocities stored in a data base were retained to 6 significant figures, surely much more accurately than the truncation error. However, this "clipping" of the velocity field led to larger errors in the discrete divergence operator. When passed to the transport code, these divergence errors produced anomalous source terms because of the fully conservative differencing. Note the nearly complete orthogonality of the two concepts involved here. One could imagine a flow solution in a 2 $\times$ 2 cell grid that possessed no useful accuracy, yet the transport solution would have negligible artificial source terms provided that the residual was negligible.

9.5 Grid Sensitivity and Word Length Sensitivity

As we have seen (in Chapter 1, Section 1.5 and elsewhere), the MM are very sensitive to cell aspect ratio, and are limited in problem size. For extremely favorable cases, such as highly stretched grids used for turbulent boundary layer calculations, it might be possible to use on the order of 1,000 points in the marching direction. In the direction transverse to the march, the limit (without going to more than double precision on usual computers) is probably about 200. Also, MM are sensitive to word length of the computer. Without too much difficulty, a reasonable upper limit is of the order of 100 $\times$ 100 grid points. MG is somewhat sensitive to the cell aspect ratio, but only in the sense of degrading iterative performance. For example, a graded mesh in both directions, such as a grid to resolve boundary layers on all four walls of a driven cavity problem, cannot be solved by MM without tortuous contrivances in patching regions together. The basic MM will just completely fail, whereas MG convergence may be seriously degraded. In extreme cases, convergence might be prevented entirely in a particular code (e.g., if ADI is used as the smoother and the cell aspect ratio is large) but generally, MG is less sensitive to the grid. MG has no inherent limitation on problem size, other than the requirement of any discrete method to resolve the small differences within the floating point precision of the method. MG is quite forgiving of short word length. (Incidentally, it is not well-known that point SOR performance is seriously degraded for very fine resolutions because of sensitivity to finding otimimum relaxation factors, and that evidence of short wordlength problems can be manifest at reasonable grid resolutions; e.g., see [19].)

MG also works best if the grid size is a highly composite multiple of 2 in each direction. If the grid is not, MG can still be made to work, but with a penalty in computer time, some penalty in convergence rate, and significant penalty in coding complexity, especially in 3-D. For this reason, almost all early MG papers used grid halving to define subgrids. There is no such restriction in MM, with the exception of the third direction for the 3-D EVP-FFT method (Chapter 5, Section 5.8).

9.6 Directionality

The MM are strongly directional, and require the user to choose a marching direction with due consideration given to cell aspect ratios and mesh dimensions. (Fundamentally, the effect is due to anisotropy of the PDE stencil, which may just as well arise from continuum equation properties as from grid properties.) MG are directional in that the convergence rate may suffer, for example, if a line SOR smoother is applied in the wrong direction in a boundary layer problem. (As described earlier, in extreme cases such as an ADI smoother with a large cell aspect ratio, directionality might prevent convergence.) Generally, MG is less sensitive than MM to directionality. However, the poor performance of classical (point smoother) MG has motivated the development of "semi-coarsening" MG algorithms [21] which pay for their improved robustness with increased operation count and storage penalty, and sensitivity to the choice (hard-wired in a code, not user-selectable) of the semi-coarsening direction. In contrast to the theme of early MG papers, it is widely recognized that there are many varieties of MG algorithms required for various problems.

9.7 Storage Penalty

For the basic MM on a single equation in 2-D, the storage penalty is "1^+", by which we mean to indicate that the storage penalty is the same size as "1" of the problems being solved, "+" some higher order terms. That is, a single 2-D array is needed for the influence coefficient matrix, plus a few 1-D work arrays. If selective double precision is needed, the storage penalty is 2^+. In 3-D, using the FFT in the third direction on a restricted class of problems (Chapter 5, Section 5.8), the storage penalty is only $1/2^+$. For classical point-solver [1] MG, the asymptotic (large, highly composite grids) storage penalty for a constant or variables separable problem is 1.5^+ in 2-D, and 1^+ in 3-D. For a general coefficient problem, it is necessary in MG to define (and possibly store) new coefficients (from the appropriate discretization) at all grid levels, which can be a chore and which requires some additional computer time. However, the present author has devised a fairly efficient scheme for this, applicable to "black box" MG codes in any dimensionality [22].

If robustness requires the semi-coarsening MG approach [21], the storage penalty is more severe: $2M^+$, when M is the fine-grid size in the semi-coarsening direction. Also, if variational methods are used to define the coarse grid operators [21] rather than the original [1] or an artificial [22] PDE, then the coarse grids require storage of a 9-point operator even if the fine-grid equation uses only a 5-point stencil. For fine grids that are not powers of two, it requires a complicated algorithm to predict the semi-coarsening MG storage required [21].

9.8 Dimensionality

Basic MM applies efficiently in 3-D only to problems that are separable in the third direction with constant coefficient even-ordered derivatives in the third direction. For this restricted class, the MM are very effective. Also, MM can be extended as an iterative 3D method but with significant loss of appeal (Chapter 5, Sections 5.5 and 5.6). MG extends readily to 3-D or higher, although the coding gets somewhat subtle. For classical point-solver MG [1], the efficiency increases significantly, and the storage penalty decreases, with increased dimensionality, a remarkable and perhaps unique attribute. The applicability to 3-D problems is possibly the greatest advantage of MG methods. With modern computer power, 3-D problems are more feasible, especially in aerodynamics. However, 2-D problems in fluid dynamics still represent a significant percentage of work. In 1993, the four issues of the ASME Journal of Fluids Engineering published 42 Computational Fluid Dynamics articles, 29 of which (69%) were exclusively 2-D. In four issues of the AIAA Journal over a similar period (9/92, 10/92, 11/93, 12/93) there were 36 CFD articles, 24 of which (67%) were 2D. About 2/3 of the work is 2-D.

If robustness requires the semi-coarsening MG approach [21] in 3-D, and line semi-coarsening is used, the storage penalty is severe: $2MN^+$, where M and N are the fine grid sizes in the two line-semi-coarsening directions. Plane relaxation should be used, and MM is a candidate (Chapter 5, Section 5.12).

9.9 Work Estimates

Work estimates for basic MM are independent of truncation error, whereas the most optimistic work estimate for MG depends on the number of iterations being a function of truncation error. (See Section 9.4 above.)

9.10 Boundary Conditions

Both MM and MG are notable for the generality of the boundary conditions, but MG are even more so. In MM, periodic boundary conditions can be applied conveniently only in the direction transverse to the marching direction (Chapter 1, Section 1.2.6). Also, tangential derivative conditions, which can arise from simple Neumann conditions applied in a locally non-orthogonal grid, can only be applied directly with one-sided differencing, due to the requirement for separability of the equation in the marching direction (Chapter 1, Section 1.2.6). Note, however, that the possible solution oscillations which can arise from centered differencing of the tangential derivative are a characteristic of the difference equations being solved, and are therefore common to MM, MG, and all solution procedures. Note also that MG can be as quirky as MM, e.g., a 2-D semi-coarsening MG algorithm [21] sometimes fails for even values of JL, even after years of algorithm development effort.

9.11 General Coefficient Problems

For a general coefficient problem, e.g., one obtained from a non-orthogonal grid or from random variation of properties, the coefficients can be actualized as arrays or regenerated when needed. Both MM and MG methods are amenable to either approach. For MM, only the final (fine grid) coefficients are needed, whereas MG requires that coefficients be generated and possibly stored for all the multigrid levels used, involving storage, arithmetic, and coding complexity penalties.

9.12 Grid Transformations

Irregular geometries can be treated by many solution methods if grid trans-formations are used to map the region to a regular (e.g., Cartesian) logical space, with subsequent change in the governing (or "hosted") equations. MM and MG are no exceptions. MM are well suited to the problem of elliptic grid generation in 2-D [23-25] because the same matrix applies to both the x and y grid equations, so that only one initialization is required. Also, boundary fitted grids tend to *help* the stability problems of MM, since the *pertinent cell aspect ratio is that in physical space.*

9.13 Irregular Logical-Space Geometry

In some applications, it is preferable to treat irregular geometries by staying in the physical space and using an irregular grid. Alternately, a hybrid approach can be used in which the equations are transformed from physical to logical space, but some irregularity is used in the logical-space grid, e.g., L- or T- shaped regions. It is relatively straight-forward to solve MM on irregular grids (Chapter 1, Section 1.2.8) although such generality in a general purpose code with many options would be difficult. However, the MG methods also appear to be difficult to apply in such cases even for a special purpose code.

9.14 Higher Order Systems

Two coupled second order equations, or a single fourth order equation, can be solved by MM (Chapter 2, Section 2.4). The method is quite attractive and practical, especially in the fluid dynamics application to the stream function-vorticity formulation and similar systems since it obviates the necessity for relaxation at no-slip boundaries. There is some degradation in operation count for initialization and in stability, but it is not serious. The problems get serious for a sixth order system, and MM appear impractical for anything larger. For MG, there appears to be a serious need for more work on multiple equation systems, which introduce new possibilities for and sensitivity to smoothing algorithms, prolongation and restriction operators, etc. It now appears clear that MG performance necessarily degrades with increasing order of the system [9,15].

9.15 Higher Order Accuracy Equations

It is disastrous to attempt to solve a higher order accuracy stencil directly using MM, but fourth order accuracy is readily obtained by deferred corrections (Chapter 2, Section 2.3). The method is fast and stable, and the iterative correction can include nonlinear iterations, truncation error estimation, etc. MG methods also seem to work best with a second order accurate smoother, and the deferred correction approach can be used within the grid transfers. Deferred corrections appear to be the method of choice for many direct and iterative solvers. (For a direct fourth order MG solution, albeit with only second order accuracy near boundaries, see [26].)

9.16 Finite Element Equations

Many finite element discretizations, say for the Poisson equation, are fourth order accurate for the homogeneous problem. Even though the problem being solved may be nonhomogeneous, the marches to establish the influence coefficient matrix are homogeneous and therefore higher order accurate and highly unstable, as noted in Chapter 2, Section 2.6. The only approach via MM would have to be through a deferred correction (Chapter 2, Section 2.3). On the other hand, MG have been applied to finite elements directly [2].

9.17 Use in Time Dependent Problems

Both MM and MG can be applied to time evolutionary problems, but the MG "philosophy" can also be applied within the time discretization itself. Also, because of the stability considerations on Helmholtz-like equations that arise from implicit time differencing, MM are best applied to slowly varying problems with large time steps. (These minimize the Helmholtz-like term and aid stability.) In contrast, MG convergence will improve for small time steps because of the improved diagonal dominance of the system.

9.18 Cell Reynolds Number Difficulties

In the MG solutions of fluid dynamics and heat transfer problems, one can encounter grid truncation difficulties such as cell Reynolds number phenomena [16] in the coarse grid cycling that are not part of the final grid; that is, the maximum cell Reynolds number could be small in the final fine mesh, but > 2 in the coarsest grid. The resulting oscillatory solutions ("wiggles") in the coarse grid can prevent convergence of the nonlinear MG methods. It is necessary to use type-dependent upwind differencing on the coarser grids, which complicates the coding. However, the present author has developed a method for achieving this sub-grid upwinding in a "black box" MG code [22].

9.19 Virtual Problems

In the present context, by "virtual problem" we mean one that can be solved on a virtual, i.e., non-actualized, grid that is larger than available computer memory. As described in Chapter 4, Section 4.4, MM can be used for such linear problems. MG could also be used in an analogous way, even for nonlinear problems, by actualizing storage only on a coarse grid. However, re-constructing the MG fine grid solution would require more work and the storage saving would be much smaller than the MM virtual problem.

9.20 MLAT and other Grid Adaptation

Brandt's rich original papers [1] explored both MG and MLAT concepts, or Multi-Level Adaptive Techniques. They are not identical. Unfortunately, Brandt's original precise terminology has been corrupted in the literature, and "multigrid" is often used not as a technical term at all, but simply to refer to any technique that utilizes more than one grid, more often in an approach closer to MLAT than to Brandt's MG.

MLAT would now be described (using terminology systematized in the finite element community [27] but equally appropriate for finite difference and finite volume methods) as "h-type adaptivity" where "h" is a measure of grid spacing, i.e., Δx in the present context. (In finite element theory, the quantitative explicit measure for "h" is not obvious, e.g., see [27].) Basically, the grid is adapted by adding more nodes (or cells, or volumes, or elements).

The point is that MM are also applicable to MLAT, as demonstrated in Chapter 3, Sections 3-7, although the method begins to take on the flavor of MM on a base grid and MG on the subgrids.

For "r-type adaptivity," in which the fixed number of nodes (or cells, or elements) are redistributed [24,25], MM has a problem because each redistribution changes the grid metrics, requiring relatively expensive re-initialization of the solver [23]. For a steady-state problem, for which one might expect 3-5 grid adaptations, MM are still appropriate and quite competitive. For highly transient problems, MM would be less efficient than MG or other iterative methods for r-type grid adaptivity.

Both "h-type" and "r-type" adaptivity are essentially the same whether applied in finite difference, finite volume, or finite element methodologies. This is not the case for "p-type adaptivity," where p represents the order of accuracy of the discretization. Increasing the order of a difference expression ($p > 2$) or a finite volume discretization involves extending the stencil beyond nearest neighbors. In contrast, increasing p in a finite element methodology involves increasing the number of nodes per element. (In my opinion, only for $p > 2$ for a second-order continuum equation is there a significant distinction between finite element methods and non-orthogonal finite difference and finite volume methods.) The essential distinction is that the *support* for the function increases with high-order finite elements, not for finite differences or finite volumes. Thus *p-type adaptivity for elements is more akin to h-type adaptivity in finite differences than it is to higher order differences.* (High order elements are appropriate for regions with strong solution structure [27] whereas higher order differences are appropriate for regions with weak solution structure [17].

In any case, since MM are not well suited to either low-order finite elements or to higher order finite differences (Chapter 2 and Chapter 9, Sections 9-15, 16), they are certainly not suited for p-type adaptivity. MG are in principal, although I know of no such application.

9.21 Vectorization, Parallelization, and Convergence Testing

MM do not vectorize well for the 9-point operator, because of the use of the tridiagonal Thomas algorithm for the march. But the LU decomposition and backsolve vectorize very well indeed (see Chapter 7), with no programming effort, on the Cray-2 and any other parallel computer with a comparably short vector half performance length. More importantly, parallelization of the march is natural, and for the "inversion" MM parallelize as well as the LU method used.

Vectorization and parallelization of MG is currently an active research area, but will at least involve considerable programming effort. MM have an elementary data structure, whereas that of MG is inherently complex, especially for grids that are not powers of two. This presumably leads to memory bank conflicts [28, p. 259] on the Cray-2 for the interpolation and prolongation steps in MG. The advantage of using successive multigrids related by a power of 2 can also add to memory bank conflicts on most machines [28, p. 260]. The subject of Vectorization was the official theme of the Second Copper Mountain Multigrid Conference [4].

Related to vectorization is the question of iterative convergence testing for MG, which is not required at all in MM. (The final residual error, which is isolated at the upper boundary in the basic MM, can be checked after the solution in a 1-D vectorizable calculation.) MG requires convergence testing in most real applications. MG research codes have been written with fixed cycles of iteration and unconditional transfers between grids, but this requires a restricted class of problems and does not appear practical for general codes used in engineering applications.

When a MM code fails, it usually fails *big*, whether failure is due to true instability or to a coding error. This is more often an advantage than a disadvantage compared to MG (and more so, other iterative methods) which can tease you on with the promise of incrementally improved performance.

9.22 Simplicity, Modularity, and Robustness

The qualities of simplicity, modularity, and robustness are desirable, but unfortunately tend to be personal judgements. I have too much experience and ego-involvement with MM to be able to evaluate these objectively. My limited experience with MG on simple elliptic problems has shown MG to be simple but quite subtle (and aesthetically appealing). The basic MM appear to me, after many years of experience, to be very straight forward with any given geometry and set of boundary conditions. For example, it is not difficult to modify a MM code for a new class of geometry such as a T-shaped region or a new class of boundary conditions such as periodic boundary conditions along only part of the boundary. But the simplicity disappears when one attempts to build a general code with many options. The complexity appears to increasc, and the code readability to decrease, exponentially with the number of options. I suspect, and have some anecdotal evidence from MG workers, that this is also the case with MG and, in fact, with any but the simplest and notoriously inefficient point iterative methods. However, if these other problem variations are required in a *stabilized* MM, the attempt to build a very general code would not be simple, and MG would seem to have an advantage.

In early papers on MG [e.g., 1] all problems appeared solvable with any of several MG algorithms. In a more recent paper [Brandt and Yavneh, 29] the authors recommend distinct MG algorithmic treatments for inflow-type regions, outflow, boundary layer, and strongly recirculating regions in order to achieve true multigrid performance. They find that the traditional (and simplest) prototype of a Navier-Stokes problem, incompressible flow in a driven cavity, "include[s] so many of these features (recirculation, boundary layers, singularities, poor approximation of some components), each of which requires its own special handling, that it is quite impossible to conclude from the results which of these is slowing down the solution process." Certainly we can fairly claim that the relative simplicity, modularity, and robustness of marching methods and multigrid methods are debatable.

9.23 Summary

An extensive and only partially subjective comparison of marching methods to multigrid methods has been given. The most significant advantage of the marching methods would seem to be the greater speed for repeat solutions, especially when tight iterative convergence or small residuals are demanded and when the equation solved is not strongly elliptic everywhere. The most significant advantage of the multigrid methods would seem to be in the application to variable coefficient problems in 3-D.

References for Chapter 9

1. A. Brandt, Multi-Level Adaptive Solutions to Boundary-Value Problems, *Mathematics of Computation*, Vol. 31, 1977, pp. 333-390.
2. R. A. Nicolaides, On the L_2 Convergence of an Algorithm for Solving Finite Element Equations, *Mathematics of Computation*, Vol. 31, 1977, pp. 892-906.
3. A. Brandt, *1984 Multigrid Guide with Applications to Fluid Dynamics*, Monograph, GMD-Studie 85, GMD-FIT, Postfach 1240, D-5205, St. Augustin 1, Germany, 1985. Also available from Dept. of Mathematics, University of Colorado at Denver, Colorado 80204-5300.
4. S. McCormick and U. Trottenberg, eds., *Proc. International Conference on Multigrid Methods*, Copper Mountain, Colo., April 1983. In *Applied Mathematics and Computation*, Special Issue on Multigrid Methods, Vol. 13, 1983.
5. S. F. McCormick, ed., *Multigrid Methods*, SIAM Frontiers in Applied Mathematics, Vol. 3, SIAM, Philadelphia, 1987.
6. S. F. McCormick, *Multilevel Adaptive Methods for Partial Differential Equations*, SIAM Frontiers in Applied Mathematics, Vol. 6, SIAM, Philadelphia, 1989.
7. S. F. McCormick, *Multilevel Projection Methods for Partial Differential Equations*, CBMS-NSF Regional Conference Series in Applied Mathematics, Vol. 62, SIAM, Philadelphia, 1989.
8. S. F. McCormick, ed., *Multigrid Methods:Theory, Applications, and Supercomputing*, Marcel Dekker, Inc., New York, 1988.
9. D. Nelson, T. Manteuffel, and S. F. McCormick, eds., *Sixth Copper Mountain Conference on Multigrid Methods*, NASA Conference Publication 3224, Vol.1 and Vol 2, 1994.
10. J. E. Dendy, *Black Box Multigrid, Journal of Computational Physics*, Vol. 48, 1982, pp. 366-386.
11. K. Stüben and U. Trottenberg, *Multigrid Methods: Fundamental Algorithms, Model Problem Analysis and Applications, Lecture Notes Math.* Vol. 960, Springer-Verlag, Berlin, 1982.
12. W. Hackbusch, *Multi-Grid Methods and Applications*, Springer Series in Computational Mathematics, Vol. 4, Springer, Berlin, 1985.
13. W. Briggs, *A Multigrid Tutorial*, SIAM, Philadelphia, 1987.
14. P. J. Roache, Additional Performance Aspects of Marching Methods for Elliptic Equations, *Numerical Heat Transfer*, Vol. 8, 1985, pp. 519-535.
15. J. Ruge, personal communication, 1994.
16. P. J. Roache, *Computational Fluid Dynamics*, rev. printing, Hermosa Publishers, Albuquerque, N. M., 1976.
17. P. J. Roache, A Method for Uniform Reporting of Grid Refinement Studies, ASME FED-Vol. 158, *Quantification of Uncertainty in Computational Fluid Dynamics*, ASME Fluids Engineering Division Summer Meeting, Washington, DC, 20-24 June 1993. I Celik, C. J. Chen, P. J. Roache, and G. Scheurer, eds., pp. 109-120. See also P. J. Roache, A Method for Uniform Reporting of Grid Refinement Studies", *Proc. 11th AIAA Computational Fluid Dynamics Conference*, July 6-9, 1993, Orlando, Fl., Part 2, pp. 1-57-1058.

18. P. J. Roache, Perspective: A Method for Uniform Reporting of Grid Refinement Studies, *ASME Journal of Fluids Engineering*, Vol. 116, Sept. 1994, pp. 405-413.
19. P. J. Roache and P. M. Knupp, Completed Richardson Extrapolation, *Communications in Applied Numerical Methods*, Vol. 9, 1993, pp. 365-374.
20. K. Salari, P. Knupp, R. Blaine, personal communication, 1993.
21. Schaffer, S., An Efficient 'Black Box' Semicoarsening Multigrid Algorithm for Two and Three Dimensional Symmetric Elliptic PDE's with Highly Varying Coefficients, *Proc. Fifth Copper Mountain Conference on Multigrid Methods*, March 31-April 5, 1991. Also, to appear in *SIAM Journal of Numerical Analysis*.
22. P. J. Roache and S. Steinberg, Application of a Single-Equation MG-FAS Solver to Elliptic Grid Generation Equations (Sub-grid and Super-grid Coefficient Generation), *Applied Mathematics and Computation*, Vol. 19, 1986, pp. 283-292.
23. P. J. Roache, Semidirect/Marching Solutions and Elliptic Grid Generation, in J. F. Thompson, ed., *Numerical Grid Generation*, Elsevier, New York, 1982, pp. 729-737.
24. J. F. Thompson, F. C. Thames, and C. W. Mastin, Automatic Numerical Generation of Body-fitted Curvilinear Coordinate System for Fields Containing Any Number of Arbitrary Two-Dimensional Bodies, *Journal of Computational Physics*, Vol. 15, 1974, pp. 299-319.
25. P. M. Knupp and S. Steinberg, *Fundamentals of Grid Generation*, CRC Press, Boca Raton, FL, 1993.
26. C. Liu and S. Liu, Multigrid Methods and High Order Finite Differences for Flow in Transition, AIAA Paper 93-3354, July 1993.
27. J. T. Oden and V. Legat, An hp Adaptive Strategy for Finite Element Approximation of the Navier-Stokes Equations, *Finite Elements in Fluids: New Trends and Applications*, Part II, Proc. VIII International Conference on Finite Elements in Fluids, Univ. Politecnica de Cataluna, Barcelona, Spain, 20-23 Sept. 1993, K. Morgan, E. Onate, J. Periaux, J. Peraire, and O. C. Zienkiewicz, Eds., pp. 32-43.
28. R. W. Hockney and C. R. Jesshope, *Parallel Computers*, Adam Hilger Ltd., Bristol, 1981.
29. A. Brandt and I. Yavneh, On Multigrid Solution of High-Reynolds Incompressible Entering Flows, *Journal of Computational Physics*, Vol. 101, No. 1, July 1992, pp. 151-164.
30. I. Yavneh, A Method for Devising Efficient Multigrid Smoothers for Complicated PDE Systems, *SIAM Journal of Scientific Computing*, Vol. 14, 1003, pp. 1437-1463.

Appendix A

MARCHING SCHEMES AND ERROR PROPAGATION FOR VARIOUS DISCRETE LAPLACIANS

A.1 Introduction

In Chapter 1, we presented a simple empirical formula, Eq. (1.2.38), for determining the maximum number of mesh points in the march direction for a target level of accuracy, for the standard 5-point discrete analog of the Laplacian operator in Cartesian coordinates. In this Appendix, we will give more detailed (and more difficult to understand) design charts for the same problem. Also, we will consider other discrete Laplacians, some of which require different marching schemes, and their stability characteristics. None of these variations appear to be very significant, but are included here for completeness. These results are all taken from our original 1970 work on the EVP marching method [1].

A.2 Standard 5-Point Laplacian

As noted in Chapter 1, the error propagation of the marching method for the standard 5-point Laplacian in cartesian coordinates has several fortunate aspects. The largest (and, therefore, most limiting) error occurs in the center of the mesh, so that we need only consider F_{ic}. Also, the effect of various conditions along the boundaries $i = 1$ and $i = I$ adjacent to the march have negligible effect on the center value even for I as small as 7, so we may neglect the I dimension and the adjacent boundary conditions as parameters of the error propagation. Finally, there is a strong effect of mesh aspect ratio $\beta = 2\Delta x/\Delta y$, which may be used to advantage. As stated in Chapter 1, the leading term in F_{ic} at J is $2(1+\alpha)^{J-2}$, where $\alpha = \beta^{-2}$. Small β thus has an adverse effect on error propagation, while large β has a favorable effect. In the limit of large β, the error propagation approaches that of the one-dimensional problem, which is merely linear in J.

The dimensionless length of practical interest in determining the applicability of the method is $(J-1)/\beta = Y/\Delta x$ (that is, the number of x-increments that we can go in the y-direction). For a unit error $E_{ic} = 1$, the value $P = \log_{10}(F_{ic})$ is plotted in Figure A-1 with $\beta = \Delta x/\Delta y$ as the parameter. The resolution level (number of significant figures) S of several computers is also shown. (The prefixes SP and DP respectively refer to single precision and double precision.) As an example of interpreting this figure, consider a Cray-1s with single precision, $S = 14.45$. For $\beta = 1$, we find $S \approx P$ at $Y/\Delta x = 20$. At this condition ($J = 21$), we may expect resolution errors in $\Psi_{i,J}$ of order unity, which is ordinarily unacceptable. But at $Y/\Delta x = 10$ with $\beta = 2$ (still $J = 21$), we find $P \approx 7.5$. The difference $S - P \approx 14.5 - 7.5 = 7$ indicates that we may expect resolution errors in $\Psi_{i,J}$ of order 10^{-7}, which is generally acceptable. For example, on a CDC 6600 in a 67×31 mesh, with $\beta = 2\frac{1}{2}$, the final resolution error was of order 10^{-6}, and each solution was obtained in 0.27 seconds. Note that for $2 \lesssim \beta < 10$, the maximum attainable $y/\Delta x$ becomes approximately independent of β and of the field size J. (But for β large enough, the error propagation is linear in J, and $y/\Delta x \rightarrow 0$.)

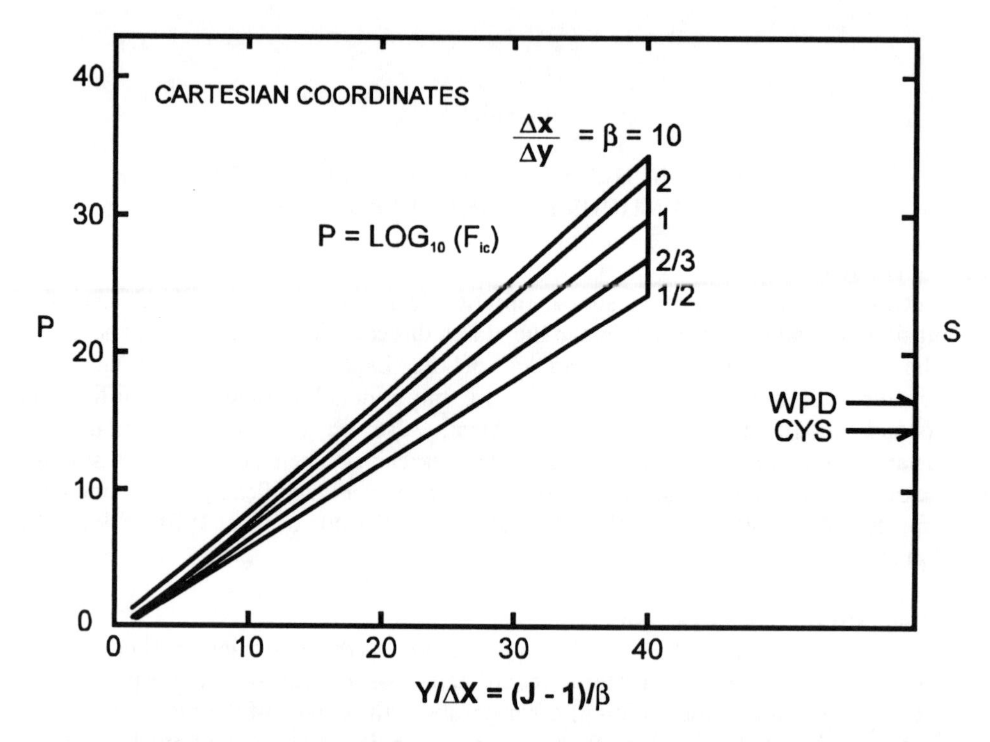

Figure A-1. Error propagation characteristics for the EVP Marching Method applied to the Poisson equation in cartesian coordinates. CYS = Cray-YMP in Single precision. WPD = Workstations and PC's (IEEE Standard) in Double precision.

A.3 Uneven Mesh

The marching method is easily applied to the uneven mesh equation for $\partial^2 \Psi / \partial y^2$ of Salvadori and Baron [2, p. 180], obtained by passing a second-order parabola through the three points. As an example of an uneven mesh, we consider a y-mesh change at j, with $(y_j - y_{j-1} = \Delta y_1)$ and $(y_{j+1} - y_j) = \Delta y_2$. Then with $\delta = \Delta y_2/\Delta y_1$ the march equation (1.2.17) becomes

$$\Psi_{i,j+1} = [\zeta_{i,j} - (\Psi_{i+1,j} + \Psi_{i-1,j} - 2\Psi_{i,j}) / \Delta x_2] \Delta y_2^2 \frac{(\gamma+1)}{2\gamma} + (1+\gamma)\Psi_{i,j} - \gamma \Psi_{i,j-1} \tag{A-1}$$

The error propagation characteristics obviously depend on the Δy schedule and cannot be presented in generality. But for the particular case of a single change at a location j*, the total error characteristic may be estimated from Figure A-1. Entering with $Y/\Delta x = (j^*-1)\Delta y_1/\Delta x$ and $\beta_1 = \Delta y_1/\Delta x$, a value P_1 at j* is read. Again entering $Y/\Delta x = (J-j^*)\Delta y_2/\Delta x$, a value $\beta_2 = \Delta y_2/\Delta x$, a value P_2 is read. The final value for the composite mesh is $P \approx P_1 + P_2$.

A.4 Other Elliptic Operators

The marching method is applicable to general elliptic operators. We briefly describe the application to some of these.

The three-dimensional Poisson equation $\nabla^2 \Psi = \zeta$ in cylindrical coordinates is

$$\frac{1}{r}\frac{\partial}{\partial r}\left[r\frac{\partial \Psi}{\partial r}\right] + \frac{1}{r^2}\frac{\partial^2 \Psi}{\partial \theta^2} + \frac{\partial^2 \Psi}{\partial z^2} = \zeta \tag{A-2}$$

The use of centered 3-point differencing on this form, without expanding the first term by the product rule for differentiation, assures that the finite-difference expression possesses the conservative property, in that the Gauss divergence relation for the continuum equation is maintained in the finite difference form.

Solving the analog of (A-2), written with $z_i = (i-1)\,\Delta_z, r_j = (j-1)\,\Delta r$ and $\theta_k = (k-1)\,\Delta\theta$, for $\Psi_{i,j+1,k}$ gives the following march equation.

$$\begin{aligned}\Psi_{i,j+1,k} &= \\ &\Big[\Delta r^2 \zeta_{i,j,k} - \alpha_j\big(\Psi_{i+1,j,k} + \Psi_{i-1,j,k} - 2\Psi_{i,j,k}\big) \\ &\quad - \alpha\big(\Psi_{i,j,k+1} + \Psi_{i,j,k-1} - 2\Psi_{i,j,k}\big)\Big]\frac{r_j}{r_{j+½}} \\ &\quad + \Psi_{i,j,k} + \frac{r_{j-½}}{r_{j+½}}\big(\Psi_{i,j,k} - \Psi_{i,j-1,k}\big)\end{aligned} \tag{A-3}$$

where

$$\begin{aligned}\alpha_j &= (\Delta r/\Delta\theta \cdot r_j)^2 \\ \alpha &= (\Delta r/\Delta z)^2\end{aligned} \tag{A-4}$$

By setting α_j or α to 0, either two-dimensional equation may be obtained. The polar coordinate case is obtained with $\alpha = 0$. The rotationally symmetric case is obtained with $\alpha_j = 0$. The physical problems that it represents (steady-state temperature distribution, electrostatic potential, etc.) are usually solved with the condition $\partial\Psi/\partial r = 0$ at $r = 0$. The error propagation characteristics for the rotational symmetric case with this boundary condition are shown in Figure A-2, in which $R = (J - 1)\,\Delta r$ is the maximum r-coordinate in the mesh. The error propagation is seen to be slightly *more favorable* than the corresponding Cartesian problem of Figure A-1.

Similarly, for the Poisson equation in spherical coordinates,

$$\frac{\partial}{\partial r}\left[r^2\frac{\partial \Psi}{\partial r}\right] + \frac{1}{\sin\theta}\frac{\partial}{\partial \theta}\left[\sin\theta\frac{\partial \Psi}{\partial \theta}\right] + \frac{1}{\sin^2\theta}\frac{\partial^2 \Psi}{\partial \varphi^2} = \zeta r^2 \tag{A-5}$$

we obtain, with $\varphi = (i-1)\Delta\varphi, r = (j-1)\Delta r$ and $\theta = (k-1)\Delta\theta$, using centered differences on the conservation form,

$$c_j = \frac{\Delta r^2}{r_{j+½}^2}$$

$$a_{j,k} = \frac{c_j}{\sin\theta_k} \Delta\theta^2$$

$$b_{j,k} = \frac{c_j}{\sin^2} \theta_k$$

$$\begin{aligned} \Psi_{i,j+1,k} = \; & c_j \Delta r^2 \zeta_{i,j,k} - a_{j,k} \left[\sin\theta_{k+½} \left(\Psi_{i,j,k+1} - \Psi_{i,j,k} \right) \right. \\ & \left. - \sin\theta_{k-½} \left(\Psi_{i,j,k} - \Psi_{i,j,k-1} \right) \right] - b_{j,k} \left[\Psi_{i+1,j,k} \right. \\ & \left. + \Psi_{i-1,j,k} - 2\Psi_{i,j,k} \right] + \left[\frac{r_{j-½}}{r_{j+½}} \right]^2 \left(\Psi_{i,j,k} - \Psi_{i,j-1,k} \right) + \Psi_{i,j,k} \end{aligned} \tag{A-6}$$

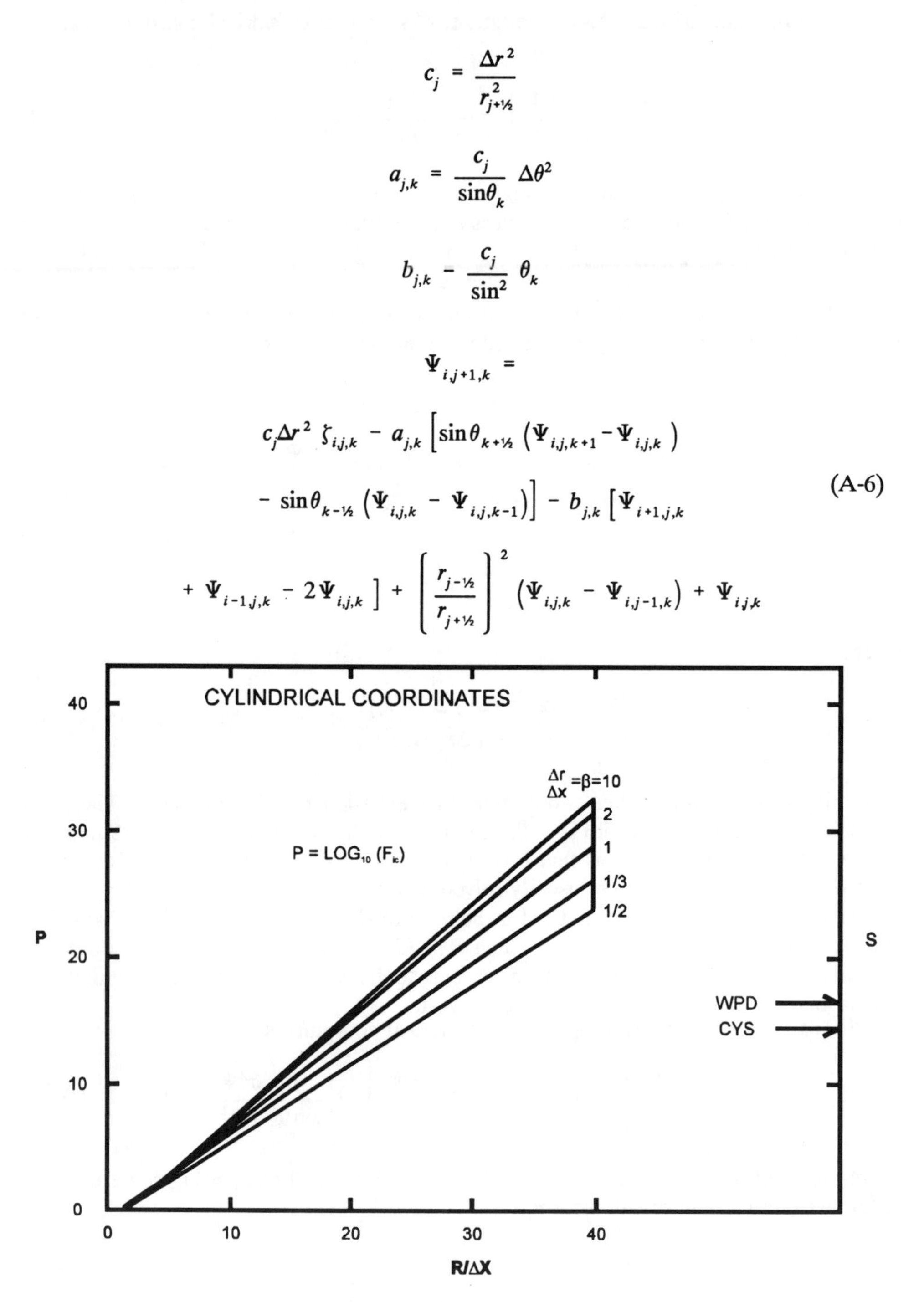

Figure A-2. Error propagation characteristics for the EVP Marching Method applied to the Poisson equation in cylindrical coordinates. CYS = Cray-YMP in Single precision. WPD = Workstations and PC's (IEEE Standard) in Double precision.

We denote by G_1^2 the elliptic operator applicable to incompressible flow in cylindrical coordinates with θ-rotational symmetry and θ-vorticity component = 0 [3, p. 131].

$$G_1^2 \Psi = \frac{\partial^2 \Psi}{\partial r^2} - \frac{1}{r}\frac{\partial \Psi}{\partial r} + \frac{\partial^2 \Psi}{\partial z^2} = \zeta \tag{A-7}$$

Centered differencing of this form, with $\alpha = (\Delta r/\Delta x)^2$, $r = (j-1)\Delta r$ and$z = (i-1)$ would give

$$\Psi_{i,j+1} = \left[\Delta r^2 \zeta_{i,j} - \alpha\left(\Psi_{i+1,j} + \Psi_{i-1,j} - 2\Psi_{i,j}\right) + 2\Psi_{i,j} - \left[1 + \frac{\Delta r}{r_j}\right]\Psi_{i,j-1}\right] \Big/ \left[1 - \frac{\Delta r}{r_j}\right] \tag{A-8}$$

This form is unacceptable near the axis. With $\Psi_{i,1} = 0$ and $\Psi'_{i,2} = E_{\ell'}$ the calculation of $\Psi'_{i,3}$ in Eq. A-8 gives a denominator at $j = 2$ of

$$\left[1 - \frac{\Delta r}{r_j}\right]_{j=2} = 1 - \frac{\Delta r}{\Delta r} = 0 \tag{A-9}$$

This situation can be remedied by using the following operator.

$$\tilde{G}_1^2 \Psi = \frac{1}{r}\frac{\partial}{\partial r}\left[r\frac{\partial \Psi}{\partial r}\right] - \frac{2}{r}\frac{\partial \Psi}{\partial r} + \frac{\partial^2 \Psi}{\partial z^2} = \zeta \tag{A-10}$$

In the continuum, $G_1^2\Psi = \tilde{G}_1^2$, but their finite difference analogs are not equal. Using the conservation form, $\tilde{G}_1^2\Psi = \zeta$ is approximated by

$$\Psi_{i,j+1} = \left\{ \left[\Delta r^2 \zeta_{i,j} - \alpha\left(\Psi_{i+1,j} + \Psi_{i-1,j} - 2\Psi_{i,j}\right)\right]\frac{r_j}{r_{j+½}} + \Psi_{i,j} + \frac{r_{j-½}}{r_{j+½}}\left(\Psi_{i,j} - \Psi_{i,j-1}\right) - \frac{\Delta r}{r_{j+½}}\Psi_{i,j-1} \right\} \Big/ \left[1 - \frac{\Delta r}{r_{j+½}}\right] \tag{A-11}$$

This conservation form causes no difficulty near the axis.

Next, we denote by G_2^2 the elliptic operator applicable to incompressible flow in spherical coordinates with φ-rotational symmetry and φ-velocity component = 0 (pg. 131 of Ref. 13).

$$G_2^2 \Psi = \frac{\partial^2 \Psi}{\partial r^2} + \frac{\sin\theta}{r^2}\frac{\partial}{\partial \theta}\left[\frac{1}{\sin\theta}\frac{\partial \Psi}{\partial \theta}\right] = \zeta \tag{A-12}$$

With $\theta = (i-1)\Delta\theta$, $r = (j-1)\Delta r$, and $\alpha = (\Delta r/\Delta\theta)^2$, centered differencing of the conservation form gives the following equation.

$$\Psi_{i,j+1} = \Delta r^2 \zeta_{i,j} - \alpha \frac{\sin\theta_i}{(r_j \Delta\theta)^2} \left[\frac{\Psi_{i+1,j} - \Psi_{i,j}}{\sin\theta_{i+\frac{1}{2}}} - \frac{\Psi_{i,j} - \Psi_{i-1,j}}{\sin\theta_{i-\frac{1}{2}}} \right] + 2\Psi_{i,j} - \Psi_{ij-1} \tag{A-13}$$

No difficulty arises near $r = 0$*(at* $j = 1$*)*, and the condition along $\theta = 0$*(at* $j = 1$*)*, is simply $\Psi_{1,j}$, from the physics of the problem and the definition of Ψ.

A.5 Other Analogs for the Laplacian

The marching method is adaptable in principle to other finite difference analogs of the Laplacian operator. Although the new error propagation characteristics are sometimes adverse, it is interesting to consider their novel aspects.

Some of these analogs are depicted in Figure A-3 for the case of $\Delta x = \Delta y = \Delta$. The schematic representation, terminology, and historical credits are taken from the classic work of Thom and Apelt [4]. The five-point analog that we have been considering was first used by Runge in 1908. It is sometimes called the "basic unit square" and is depicted in Figure A-3a. This schematic applied to the Poisson equation $\nabla^2\Psi = \zeta$ represents the following equation, which is second-order accurate in Δ.

$$\Psi_{i+1,j} + \Psi_{i-1,j} + \Psi_{i,j+1} + \Psi_{i,j-1} - 4\Psi_{i,j} = \Delta^2 \zeta_{i,j} \tag{A-14}$$

Since the Laplacian operator is invariant to a coordinate transformation, it may be expressed in coordinates rotated 45° with respect to the mesh, in which case the spacing between points becomes $\sqrt{2}\,\Delta$. The resultant "diagonal unit square" operator is shown in Figure A-3b, which represents

$$\Psi_{i+1,j+1} + \Psi_{i+1,j-1} + \Psi_{i-1,j+1} + \Psi_{i-1,j-1} - 4\Psi_{i,j} = 2\Delta^2 \zeta_{i,j} \tag{A-15}$$

which is second-order accurate in $\sqrt{2}\,\Delta$.

A possible march scheme for Eq. A-15 is shown in Figure A-4. The arrows indicate the computational sequence, with the head of the arrow being the new calculated point $\Psi_{i+1,j+1}$, obtained from Eq. A-15. Note that the initial and final error vectors E and F each extend along *two* boundaries in this scheme. Unlike the basic unit square operator, the diagonal operator has error propagation characteristics that depend strongly on the mesh dimension I in the direction transverse to the march direction. These characteristics are plotted in Figure A-5. For $I = 11$, the diagonal operator is much better than the basic unit square, but for $I = 51$, the advantage is reversed.

$$\begin{bmatrix} & 1 & \\ 1 & -4 & 1 \\ & 1 & \end{bmatrix}$$

a. BASIC UNIT SQUARE

$$\frac{1}{2}\begin{bmatrix} 1 & & 1 \\ & -4 & \\ 1 & & 1 \end{bmatrix}$$

b. DIAGONAL UNIT SQUARE

$$\frac{1}{4}\begin{bmatrix} 1 & 2 & 1 \\ 2 & -12 & 2 \\ 1 & 2 & 1 \end{bmatrix}$$

c. THE "12" FORMULA

$$\frac{1}{6}\begin{bmatrix} 1 & 4 & 1 \\ 4 & -20 & 4 \\ 1 & 4 & 1 \end{bmatrix}$$

d. THE "20" FORMULA

Figure A-3. Stencil of various finite difference analogs of the Laplacian operator in cartesian coordinates with $\Delta x = \Delta y$.

The two most popular 9-point formulae are Thom's "12" formula, Figure A-3c, and Bickley's "20" formula, Figure A-3d. The "12" formula is

$$\begin{aligned} &2\left(\Psi_{i+1,j} + \Psi_{i-1,j} + \Psi_{i,j+1} + \Psi_{i,j-1}\right) \\ &+ \left(\Psi_{i+1,j+1} + \Psi_{i+1,j-1} + \Psi_{i-1,j+1}\right. \\ &\left. + \Psi_{i-1,j-1}\right) - 12\Psi_{i,j} = 4\Delta^2 \zeta_{i,j} \end{aligned} \tag{A-16}$$

and the "20" formula is

$$\begin{aligned} &4\left(\Psi_{i+1,j} + \Psi_{i-1,j} + \Psi_{i,j+1} + \Psi_{i,j-1}\right) \\ &+ \left(\Psi_{i+1,j+1} + \Psi_{i+1,j-1} + \Psi_{i-1,j+1} + \Psi_{i-1,j-1}\right) \\ &- 20\,\Psi_{i,j} = 6\Delta^2 \zeta_{i,j} \end{aligned} \tag{A-17}$$

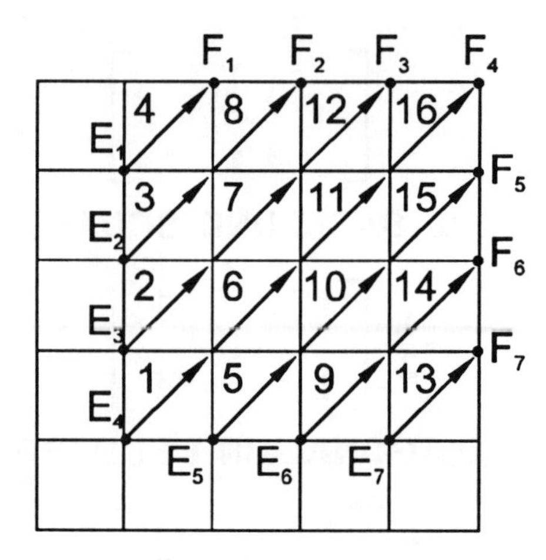

Figure A-4. March scheme for the 5-point diagonal unit square operator and the explicit march sheme for the 9-point operators.

Equation A-16 is fourth order accurate only if

$$1/12\left(\nabla^2\nabla^2 + \partial^4/\partial x^2\partial y^2\right)\Psi = 0\left(\Delta^2\right) \tag{A-18}$$

or smaller and (A-17) only if

$$1/12\,\nabla^2\nabla^2\Psi = 0\left(\Delta^2\right). \tag{A-19}$$

It can be shown that this condition will be met in fluid dynamics problems only in very low Reynolds-number flows. The results of Jenssen and Straede [5] indicate that these 9-point analogs are not generally more accurate than the 5-point formula, Eq. A-14.

Either 9-point formula could be marched implicitly using a tridiagonal solver for each j-line, as in Section 1.3.6 for cross-derivatives. Alternately, with the initial guess extending along two adjacent boundaries, it may be marched out using the same explicit diagonal march scheme depicted in Figure A-4, but using Eq. A-16 or A-17 to solve for $\Psi_{i+1,j+1}$. The implicit method is more time-consuming to march than the explicit method, since the equivalent of two 9-point Richardson iterations are replaced by the equivalent of two sweeps (one complete cycle) of a 9-point ADI iteration. However, the LU solution is more expensive since E and F are (roughly speaking) twice as long. In spite of the fact that the adjacent boundary conditions now affect every point calculation in an elliptic manner, the error propagation characteristics are *adversely* affected. In Figure A-5, we have plotted P for the 12-point analog using an implicit march, with I = *11.* It is clearly much worse than the basic unit square operator.

The marching method is readily adapted to fourth-order analogs to partial derivatives, but not to the entire Laplacian. As an example, we consider the following analog, one of several fourth-order methods given by Jenssen and Straede [5].

$$\frac{\partial^2\Psi}{\partial x^2} \simeq \frac{\delta^2\Psi}{\delta x^2} = \left[-\Psi_{i+2,j} + 16\,\Psi_{i+1,j} - 30\,\Psi_{i,j} + 16\,\Psi_{i-1,j} - \Psi_{i-2,j}\right]/12\,\Delta x^2 \tag{A-20}$$

12I(11)	"12" FORMULA, IMPLICIT MARCH, I = 11
12E(11)	"12" FORMULA, EXPLICIT MARCH, I = 11
5	5 POINT, BASIC UNIT SQUARE, I LARGE
5D(11)	5 POINT, DIAGONAL UNIT SQUARE, I = 11
5D(51)	5 POINT, DIAGONAL UNIT SQUARE, I = 51
3-5	3 POINT ANALOGUE FOR $\delta^2\psi/\delta y^2$, 5 POINT ANALOGUE FOR $\delta^2\psi/\delta x^2$, I LARGE

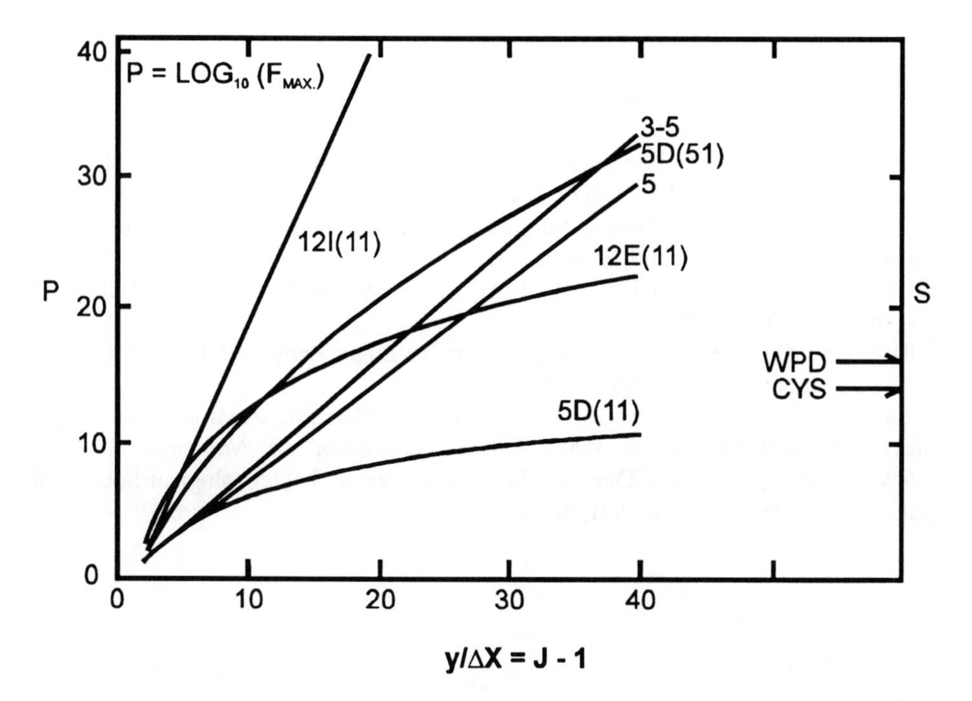

Figure A-5. Error propagation characteristics for the EVP Marching Method applied to the Poisson equation in cartesian coordinates with various analogs for the Laplacian operator, with $\Delta x = \Delta y$. CYS = Cray-YMP in Single precision. WPD = Workstations and PC's (IEEE Standard) in Double precision.

Using the 3-point second-order accurate analog for $\partial^2 \Psi / \partial y^2$, the march can proceed in the j-direction, using

$$\Psi_{i,j+1} = \Delta y^2 \left[\zeta_{i,j} - \frac{\delta^2 \Psi}{\delta x^2} \right] + 2\Psi_{i,j} - \Psi_{i,j-1} \tag{A-21}$$

Near the adjacent boundaries, at $i = 2$ and $i = I - 1$, it is necessary to revert to a 3-point analog for $\delta^2 \Psi / \delta x^2$, since δ_{i-2} and δ_{i+2}, respectively, are not available. Note that this mixed 3-point, 5-point analog, unlike those depicted in Figure A-3, is adaptable to the non-square mesh with $\Delta x \neq \Delta y$.

The error propagation characteristics for this mixed 3-point, 5-point method are shown in Figure A-5 for $\Delta x = \Delta y$. P is only 12 percent higher than the P for the basic unit square operator. This may possibly be compensated in a particular problem by taking advantage of the fourth order accuracy in the x-direction to increase Δx and therefore, β, which will reduce P (see Figure A-1).

The same device is applicable to the n-dimensional problem, as long as the 3-point analog is used in the march direction, but difficulties arise if we try to use a 5-point expression like (A-18) along the march direction. As in Chapter 2, the length of the initial and final error vectors is doubled, with bad effects on the operation count and storage penalty. More importantly, the 4-th order accuracy, deferred corrections or Richardson extrapolation are required, as in Chapter 2.

References for Appendix A

1. P. J. Roache, A Direct Method for the Discretized Poisson Equation, Sandia National Laboratories, Report SC-RR-70-579, Albuquerque, NM, February 1971.
2. M. G. Salvadori and M. L. Baron, *Numerical Methods in Engineering*, Prentice-Hall, Englewood Cliffs, NJ, Second Edition, 1961.
3. R. B. Bird, W. E. Stewart, and E. N. Lightfoot, *Transport Phenomena*, John Wiley and Sons, Inc., New York, 1960.
4. A. Thom and C. J. Apelt, *Field Computations in Engineering and Physics*, D. Van Nostrand Company, Ltd., London, 1961.
5. D. Jenssen and J. Straede, The Accuracy of Finite Difference Analogues of Simple Differential Operators, *Proc. WMO/IUGG Symposium on Numerical Weather Prediction*, Tokyo, Nov. 26 - Dec. 4, 1968. Published by Meteorological Society of Japan, March 1969. Chapter VII, pp. 59-78.

Appendix B

TRIDIAGONAL ALGORITHM FOR PERIODIC BOUNDARY CONDITIONS

B.1 Introduction

When 2D marching methods are applied to a 9-point operator (Section 1.3.6), the march proceeds one line at a time, as in line SOR. For a march in the $+y$ direction, a tridiagonal matrix solver is used to obtain new $\Psi_{i,j+1}$ for all i in the row. If the boundary conditions in x, transverse to the march direction, are periodic, this requires a tridiagonal algorithm for periodic boundary conditions.

In [1], Temperton reviewed methods for solving cyclic tridiagonal systems and presented a new method denoted therein as Algorithm 4. The emphasis in [1] was on solving symmetric circulant problems for many repeat solution. The present appendix complements [1] by considering the general coefficient case, which is required if we are solving a general coefficient 2D elliptic operator by marching methods.

The new algorithm presented herein is not a marching method, but it is still based on the simple but powerful concept of influence coefficients. (Note that the concept of "influence coefficients" is closely related to that of superposition, and both depend on linearity, but there is a difference in the detailed algorithms.) It can be used with any algorithm for non-periodic tridiagonal systems to solve the periodic problem. It has virtually no round-off error beyond that of the strictly tridiagonal algorithm, and is applicable to general coefficient problems. (This algorithm can also be used to "complete" the Algorithm 4 of [1] for general coefficient systems.)

B.2 Problem Statement

We consider algorithms for the solution of the system $A\underset{\sim}{x} = \underset{\sim}{d}$, where A is an n x n cyclic tridiagonal matrix of the following form.

$$A\underset{\sim}{x} = \begin{bmatrix} b1 & c1 & & & & & a1 \\ a2 & b2 & c2 & & & & \\ & a3 & b3 & c3 & & & \\ & \cdot & \cdot & \cdot & & & \\ & \cdot & \cdot & \cdot & & & \\ & \cdot & \cdot & \cdot & & & \\ & & & an-1 & bn-1 & cn-1 & \\ cn & & & an & bn & & \end{bmatrix} \begin{bmatrix} x1 \\ x2 \\ x3 \\ \cdot \\ \cdot \\ \cdot \\ xn-1 \\ xn \end{bmatrix} = \begin{bmatrix} d1 \\ d2 \\ d3 \\ \cdot \\ \cdot \\ \cdot \\ dn-1 \\ dn \end{bmatrix} \quad \text{(B-1)}$$

B.3 "Algorithm 4"

Algorithm 4 of [1] is based on the simple idea that if we knew the value of x_1 in Eq. (B-1), we would then obtain the order $(n\text{-}1)$ reduced system S,

$$S\,\underset{\sim}{x} = d' \tag{B-2a}$$

or

$$\begin{bmatrix} b_2 & c_2 & & & \\ a_3 & b_3 & c_3 & & \\ & a_3 & b_3 & c_3 & \\ & \cdot & \cdot & \cdot & \\ & \cdot & \cdot & \cdot & \\ & \cdot & \cdot & \cdot & \\ & a_{n-1} & b_{n-1} & c_{n-1} & \\ & a_n & b_n & c_n & \end{bmatrix} \begin{bmatrix} x_2 \\ x_3 \\ \cdot \\ \cdot \\ \cdot \\ x_{n-1} \\ x_n \end{bmatrix} = \begin{bmatrix} d_2 - a_2 x_1 \\ d_3 \\ \cdot \\ \cdot \\ \cdot \\ d_{n-1} \\ d_n - c_n x_1 \end{bmatrix} \tag{B-2b}$$

which can then be readily solved using the standard (Thomas) tridiagonal algorithm. Let $\underset{\sim}{z}=(z_1,z_2,...z_n)$ be the first row of the inverse matrix A^{-1}. Then x_1 is given by the scalar product

$$x_1 = \sum_{i=1}^{n} z_i d_1 \tag{B-3}$$

In [1], the algebraic solution for $\underset{\sim}{z}$ was given explicitly for the case of A symmetric circulant, i.e. for a_i, b_i, c_i constant and $a_i = b_i$. For the more general case the method for determining z was left open, assuming that whatever cost involved would be amortized over many repeat solutions, i.e. solutions with A fixed but varying $\underset{\sim}{d}$. For example, the methods of Ahlberg et al. [2] or Evans and Atkinson [3] could be used.

B.4 The Influence Coefficient "Algorithm 5"

The new influence coefficient algorithm, which we denote as Algorithm 5, is also based on the idea of knowing the value of x_1 in Eq. (B-1), but instead of solving for it from Eq. (B-3), we simply guess it, and evaluate the residual at $i = 1$. Two such guesses and solutions would determine the linear relation between the residual and x_1, and a third "guess" is chosen to drive the residual to zero.

This is the same simple concept of "influence coefficient" used in Sections 1.2.1 and 1.2.2 to derive the marching method. Note, however, that no "march" necessarily is involved here. The solution with a guessed value at $i = 1$ could indeed be obtained by the 1D marching method of Section 1.2.1, but it might just as well be obtained by the Thomas algorithm in any of its variations, or even by an iterative method. (We base our operation counts on the use of the Thomas algorithm for the non-periodic problem.)

The operation count and the step-by-step algorithm are given in terms of initiation and repeat solutions, with the operation count for the system S of Eq. (B-2) being $2(n-1)$ multiplications and additions for initiation, and $3(n-1)$ for repeat solutions.

The residual at $i = 1$, r_1, is evaluated for a vector $\underset{\sim}{x}$ as

$$r_1 = a_1 x_n + b_1 x_1 + c_1 x_2 - d_1 \tag{B-4}$$

By the linearity of the problem, we have

$$r_1 = s \cdot x_1 + v \tag{B-5}$$

From consideration of the principle of superposition, it is clear that s is determined entirely by the matrix A, whereas v depends on A and the source term $\underset{\sim}{d}$ as well.

To determine the coefficients s and v in Eqn. (B-5), consider the homogeneous equation $\underset{\sim}{A}x = 0$. Since $\underset{\sim}{A}$ is nonsingular, the solution is $x = 0$. Therefore, the residual r_1 will be zero when $x_1 = 0$. But then B-5 requires $v = 0$. If instead one guesses $x_1 = 1$, then $r_1 = s \cdot x_1 + v = s \cdot 1 + 0 = s$. Hence, if the reduced system (B-2a) is solved with $x_1 = 1$, then the resultant residual will determine the coefficient s (which as previously stated is independent of the source term d). Now consider the inhomogeneous equation $\underset{\sim}{A}x = d$ that we wish to solve. The residual for this equation is of the form $r_1 = s \cdot x_1 + v'$. To determine v', guess $x_1 = 0$ and solve the corresponding reduced system to get the residual r_1. In this case, $r_1 = v'$ so v' has now been determined. Finally, one solves the original system $\underset{\sim}{A}x = d$ by setting $x_1 = -v/s$ (so that $r_1 = 0$) and solving the reduced system once more.

For repeat solutions, we set $x_1^0 = 0$, and with the given source term $\underset{\sim}{d}$, obtain a repeat solution of S. The residual r_1^0 is determined from Eq. (B-4). From Eq. (B-5) with $x_1^0 = 0$, we have solved $v = r_1^0$. With the value of s obtained in the initiation, we now drive the residual $r_1^1 = 0$ in Eq. (B-5) by solving for $x_1^1 = -v/s$. With this correct value of x_1, a second solution of S gives the entire solution. A repeat solution, thus, requires two repeat solutions of S (which has already been initialized during the initiation of the cyclic solution).

B.5 Completed "Algorithm 4"

The above algorithm can now be used to complete the initiation of Algorithm 4 of [1] by providing the solution for $\underset{\sim}{z}$ = first row of A^{-1}. This is obtained by the solution of the system

$$A^T \underset{\sim}{z} = (1,0,0,...0)^T \tag{B-6}$$

Since the transpose matrix A^T is also cyclic tridiagonal, Algorithm 4 is applicable, but a further simplification accrues in the initiation procedure because of the special source term in Eq. (B-6). Without bothering with the homogeneous solution, setting $z_1^0 = 0$ in Eq. (B-6) gives the zero solution $\underset{\sim}{z}_1 = 0$, from which the residual r_1^0 can be calculated immediately from Eq. (B-4) as $r_1^0 = -1 = v$. Setting $z_1^1 = 1$ and solving S^T gives the vector $\bar{x}^1$ from which r_1^1 is calculated by Eq. (B-4). Solving for s from Eq. (B-5) gives $s = 1 + r_1^1$, so that the final value of z_1 is

$$z_1 = 1/(1 + r_1^1). \tag{B-7}$$

A second repeat solution of S^T now gives the rest of z. This part of the initiation requires the initiation of S^T, whereas repeat solutions will also require S to have been initialized. Thus, complete initialization time requires one initiation and two repeat solutions of S^T, plus one initiation of S. (A special solver could be used to save the dummy additions of the source term in Eq. (B-6), but only general-purpose solvers have been used herein).

With the vector z determined and stored in the initiation procedure, repeat solutions require n multiplications and additions to evaluate x_I from Eq. (B-3), plus one repeat solution of system S.

When LU decomposition is used to solve a banded system, the same initiation can be used for both the original matrix and its transpose [4]. In the present case of a tridiagonal banded system, the Thomas algorithm can also be arranged for initiation and repeat solutions, but the total is slightly more work than a complete solve by the original algorithm, and the additional 2D storage for the marching problem (Section 1.3.6) is expensive.

B.6 Operation Counts and Other Comparisons

The operation count for the standard (Thomas) algorithm for simple non-cyclic tridiagonal system with general coefficients is 2 multiplies and additions per mesh point for initiation and 3 for repeats; for the constant coefficient case, the count is 1 for initiation and 2 for repeats. Using these values, Table B-1 was compiled from the above description of the algorithms. On this basis, it is seen that Algorithm 5 is faster if a new coefficient system is to be solved each time, as required in the GEM codes (Chapter 6; see also [5,6].) For three or more repeat solutions, the completed Algorithm 4 would be faster. Also to be considered are the facts that the completed Algorithm 4 requires one additional vector storage for z (a consideration for repeat solutions of the 2D marching methods), and that the summation in Eq. (B-3) adds somewhat to the round-off error. Both algorithms will generalize to higher order systems. For example, the bi-tridiagonal cyclic system can be solved following the development for 4th order marching method in Section 2.4. One initiation and three repeat solutions of the simple bi-tridiagonal system (using a banded LU decomposition of other algorithm) establishes the linear relation between the dependent variables x_I and y_I and the two residuals, and a fourth repeat solution establishes the final cyclic solution with zero residuals.

Timing tests for codes utilizing these algorithms are presented in Table B-2. As noted previously, the absolute times on an old computer are not relevant, but the relative times are significant. The results generally bear out the operation counts. The round-off errors of both are excellent. The method of Ahlberg et al. [2] may have a slightly better operation count [7].

	Initiation	Repeat	New Coefficient Solution
SIMPLE TRIDIAGONAL SYSTEM			
Constant Coefficient	1	2	3
General Coefficient	2	3	5
PERIODIC TRIDIAGONAL SYSTEM, ALGORITHM 5			
Constant Coefficient	3	4	7
General Coefficient	5	6	11
PERIODIC TRIDIAGONAL SYSTEM, COMPLETED ALGORITHM 4			
Constant Coefficient	6	2.5	8.5
General Coefficient	10	4	14

Table B-1. Approximate operation counts for the algorithms considered. The numbers shown are the approximate multiplications and additions per mesh point, assuming n mesh points and n large.

(initiation time, repeat time)

n	Algorithm 5	Completed Algorithm 4
10	.59 , .78	1.30 , .43
100	3.79 , 4.33	9.49 , 2.70
1000	35.6 , 40.1	91.60 , 25.3

(maximum error, residual at I = 1)

n	Algorithm 5	Completed Algorithm 4
10	.023 (-.040)	0.23 (-.051)
100	1.07 (-.017)	1.07 (-.418)
1000	23.1 (-.17)	23.1 (-1.86)

Table B-2. Execution times and errors for the algorithms considered. Times are averaged over 100 subroutine calls and are given in milliseconds on a CDC 6600 for FORTRAN IV codes with Level 2 optimization. Errors are the residual at $i=1$ from Eq. (4) and the mean maximum errors. 100 pseudo-random solution vectors x were generated in the range $0 < x_i < 10$ and coefficients $\tilde{a}, \tilde{b}, \tilde{c}$ in the same range. The corresponding source terms were computed from $\tilde{d} = A\tilde{x}$. The algorithms were used to compute solutions $\hat{x}$ and the error $\max_i |\tilde{x}_i - \hat{x}_i|$ was averaged over the 100 cases. This averaged maximum error and the residual are displayed in units of 10^{-13}.

References for Appendix B

1. C. Temperton, Algorithms for the Solution of Cyclic Tridiagonal Systems, *Journal of Computational Physics*, Vol. 19, 1975, pp. 317-323.
2. H. H. Ahlberg, E. N. Nilson and J. L. Walsh, *The Theory of Splines and Their Applications*, Academic Press, New York, 1967, p. 15.
3. D. J. Evans and L. V. Atkinson, *Computing Journal*, Vol. 13, 1970, p. 323.
4. J. J. Dongarra, C. B. Moler, J. R. Bunch, and G. W. Stewart, *LINPACK User's Guide*, Society for Industrial and Applied Mathematics, Philadelphia, 1979.
5. P. J. Roache, GEM Solutions of Elliptic and Mixed Problems with Non-Separable 5- and 9-Point Operators, in *Proc. Elliptic Problem Solver Conference,* Santa Fe, NM, June 30-July 2, 1980, M. Schultz, ed., Academic Press, NY, pp. 399-403.
6. P. J. Roache, Performance of the GEM codes on Nonseparable 5- and 9-Point Operators, *Numerical Heat Transfer*, Vol. 4, 1981, pp. 395-408.
7. C. Temperton, personal communication, 1976.

Appendix C

GAUSS ELIMINATION AS A DIRECT SOLVER

C.1 Introduction

Gaussian elimination (abbreviated GE in this appendix) traditionally has not been used by finite difference practitioners for solutions of large multidimensional problems (with some exceptions, e.g. [1-7]. Their three objections have been poor round-off error, speed, and storage penalty. It has been used by finite element practioners quite extensively. They have tolerated the objectional aspects for several reasons. Historically, FEM practitioners typically used less resolution than FDM, not so much because of higher accuracy of FEM, as has often been claimed, but because FEM practioners have usually worked on problems requiring less resolution, and have been satisfied with less accuracy and with less efficiency. Also, FEM practioners know the characteristics of GE better.

While not at all competitive with marching methods (when the latter work), Gaussian elimination actually has a place in FDM solutions, in spite of previous claims to the contrary by myself [8] and others. From a historical perspective, all three objections have been mollified.

C.2 Round-Off Error

As late as the early 1970's, some engineering textbooks on numerical analysis were still claiming that the practical upper limit for GE was about 40 simultaneous linear equations. While this "folklore" undoubtedly was based on some realistic experience, that experience was apparently based on ill-conditioned problems, and on the use of GE without pivoting. It is now recognized that GE with at least partial pivoting (e.g. the LINPACK [9] or LAPACK codes [10]) can solve some very large systems without losing the solution in round-off error.

C.3 Speed

The speed of GE is objectionable, but not as much as previously thought. GE gives a very poor operation count for a full matrix (as shown below) and this is required for completely non-structured matrices such as those arising from some FEM discretizations of from FDM or least-squares discretizations with very large stencils. But for FDM or low-order FEM on a logically rectangular grid, the resulting matrix is simply banded, and the speed improves significantly. Furthermore, it is possible to save the LU decomposition of the matrix (provided that the grid coefficients do not change), rather than re-initializing each call, and obtain much higher speed for repeat solutions. For repeat solutions in a 2D grid, banded GE is comparable in speed to SOR, and has the added advantages of not requiring a determination (by analysis or numerical search) of optimum relaxation factors, of not being dependent on grid parameters or stretching, and of not requiring iterative convergence criteria. Also, it readily handles the expanded band-width system that results from a Newton-Raphson treatment of nonlinearities, e.g. [4-6].

C.4 Storage Penalty

Finally, the storage penalty is reduced by the use of a banded solver, and is not as important a consideration on modern computers with larger memory.

C.5 Operation Count and Storage Penalty for 2D

In this section, we will retain only the highest order terms in the operation counts. Consider the 5-point operator in an IL x JL mesh. The number of unknowns is

$$n = IL \cdot JL. \tag{C-1}$$

If no account is taken of the banded matrix structure, but rather a full matrix GE elimination routine is used, then Eq. (1.2.26) gives ~ $n^3/3$ multiplies and adds, or

$$\theta_{init,\, full\ 2D} = 2n^3/3 \tag{C-2}$$

giving

$$\theta_{init,\, full\ 2D} = \tfrac{2}{3}\, IL^3 JL^3 \tag{C-3}$$

or for IL = JL = N,

$$\theta_{init,\, full\ 2D} = \tfrac{2}{3}\, N^6. \tag{C-4}$$

The storage penalty is a matrix of order N, or $N^2 = IL^2JL^2$ storage locations, compared to the 2D problem size of IL x JL. That is, full matrix GE requires the equivalent of N^2 2D arrays for an N x N problem. It is easy to see how GE got a bad name.

The evaluation is more optimistic when advantage is taken of the fact that the matrix is not full, but banded. The matrix structure for a 2D 5-point operator is shown in Figure C-1. The structure is "block tridiagonal", and (when properly set up) the bandwidth is determined by the smaller of IL and JL. For concreteness, we demonstrate with IL < JL, specifically the small problem of IL = 7 and JL = 8, for 56 equations. We also assume Neuman (2-point gradient) boundary conditions along the top and bottom (j = 1 and j = JL) and Dirichlet boundary conditions along the sides (i = 1 and i = IL) giving diagonal entries there. (If Dirichlet boundary conditions also were used along j = 1 and j = JL, the first 8 and last 8 entries would be diagonal, in which case these known values could be eliminated easily from the matrix problem by preprocessing, reducing the size of the system matrix by 16 elements, to 40 equations. We do not consider here eliminating other Dirichlet boundary equations, for the sake of generality.)

We use the notation of Chapter 4, defining band width

$$W = 1 + 2B \tag{C-5}$$

where B = 0 for a diagonal matrix ($c_{ij} = 0$ for $i \neq j$), B = 1 for tridiagonal, etc. Inspection shows

$$B_{2D} = IL + 1 \tag{C-6}$$

for the natural (lexigraphic) ordering for the 5-point operator, and

$$B_{2D} = IL + 2 \tag{C-7}$$

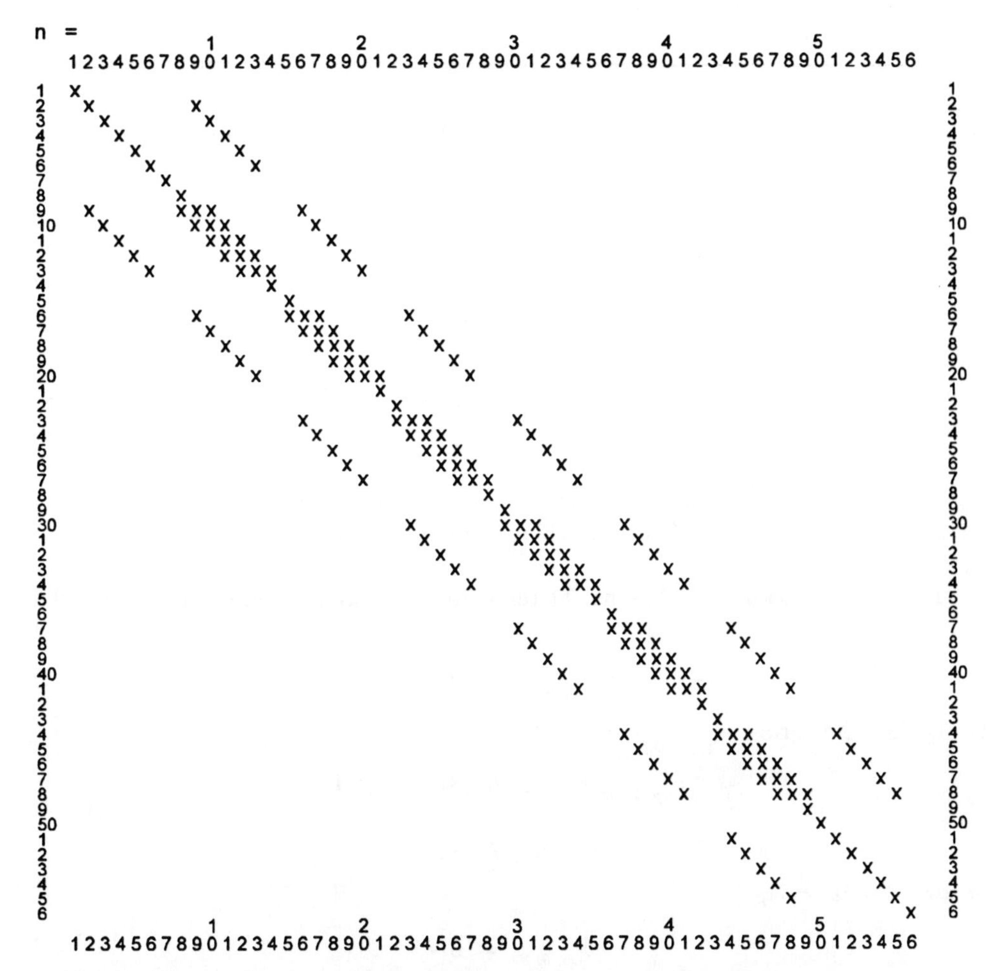

Figure C-1. Matrix structure (incidence matrix) for a 5-point operator on a 2D grid. Neuman (2-point gradient) boundary conditions are assumed along the top and bottom, j = 1 and ji = JL. Dirichlet boundary conditions are assumed along the sides, i = 1 and i = IL giving diagonal entries. IL = 7, JL = 8. Lexical ordering for the (i, j) array elements; matrix element number $n = i + IL \bullet (j - 1)$.

for the 9-point operator. (The use of the 9-point operator would affect the operation counts and storage penalties given here relatively little.)

The operation count for initiation *with* pivoting (Chapter 4, Section 4.3) gives $nB\ (2B + 1)$ multiplies and additions, or

$$\theta_{init,B,2D} = 2nB\,(2B + 1). \tag{C-8}$$

(This can be reduced to $2nB(B + 1)$ if pivoting is not required, e.g. when the marching method is applied to the Poisson equation.) As the band width increases towards a full matrix, $B \rightarrow n/2$, and Eq. (C-8) approaches

$$\theta_{init,B,2D} \simeq n^3 \tag{C-9}$$

which is larger than Eq. (C-2) for a full matrix. In the cases of interest for 2D grids, $B << n/2$. Using Eq. (C-6) gives

$$\begin{aligned}\theta_{init,B,2D} &\simeq 2IL \cdot JL(IL+1)(2(IL+1)+1) \\ &\simeq 4IL^3\, JL\end{aligned} \tag{C-10}$$

or for IL = JL = N,

$$\theta_{init,B,2D} \simeq 4N^4. \tag{C-11}$$

This is N times greater than the initiation for the marching method from Eq. (1.2.32a); for representative siza problems, marching methods initialize 1-2 orders of magnitude faster.

For repeat "backsolve" solutions, the full matrix operation count of $2n^2$ is reduced to

$$\theta_{rep,\,B,\,2D} = 2n(3B+1). \tag{C-12}$$

Using Eq. (C-6) gives

$$\begin{aligned}\theta_{rep,\,B,\,2D} &= 2IL \cdot JL(3(IL+1)+1) \\ &\simeq 6IL^2 JL\end{aligned} \tag{C-13}$$

or for IL = JL = N,

$$\theta_{rep,\,B,\,2D} \simeq 6N^3 \tag{C-14}$$

The operation count for SOR (from Chapter 1, Section 1.2.4) is $0(N^3 \ln N)$, so Eq. (C-14) indicates that banded GE for repeat solutions is at least asymptotically faster than SOR for initial solutions in 2D. Compared to marching methods in Eq. (1.2.32b), banded GE repeat solutions are 3/7 N slower, or more than an order of magnitude for N > 24.

The storage penalty is severe. The banded GE codes in LINPACK [9] require $(3B+1)$ n storage, compared to $2n$ for the problem definition. Using Eq. (C-6), this gives a storage penalty equivalent to $(3IL+4)$ 2D arrays, compared to $1+$ for the marching methods and 0 for SOR. For reasonable array sizes, this requires a large memory computer. Note, however, an ameliorating factor; when using non-orthogonal coordinates, all the information required is in the GE solver, so the metric information of the grids need not be stored.

C.6 Operation Count and Storage Penalty for 3D

Consider the 7-point 3D operator in an IL x JL x KL mesh. The number of unknowns is

$$n = IL \cdot JL \cdot KL. \tag{C-15}$$

If no account is taken of the banded matrix structure, but a full matrix GE is used, we again have

$$\theta_{init,\, full\ 3D} \simeq 2n^3/3$$
$$= \tfrac{2}{3}\, IL^3 JL^3 K^3 \tag{C-16}$$

or for IL = JL = KL = N,

$$\theta_{init,\, full\ 3D} \simeq \tfrac{2}{3}\, N^9. \tag{C-17}$$

The storage penalty is again a matrix of order n, or $n^2 = IL^2\, JL^2\, KL^2$ storage locations, compared to the 3D problem size of IL x JL x KL. That is, full matrix GE requires the equivalent of N^3 3D arrays for an N x N x N problem.

The banded matrix has band width $w = 1+2B$ where, in 3D,

$$B_{3D} = IL * JL + 1 \tag{C-18}$$

for the natural (lexigraphic) ordering of the 7-point operator. (If cross-derivatives are included, B = IL * JL + 2.) Using this equation in Eq. (C-8) gives

$$\theta_{init,\, B,\, 3D} = 2IL \cdot JL \cdot KL \cdot (IL \cdot JL + 1)(2(IL \cdot JL + 1)$$
$$\simeq 4IL^3\, JL^3\, KL \tag{C-19}$$

or for $IL = JL = KL = N$,

$$\theta_{init,\, B,\, 3D} \simeq 4N^7. \tag{C-20}$$

For repeat solutions, using Eq. (C-18) in Eq. (C-12) gives

$$\theta_{rep,\, B,\, 3D} = 2IL \cdot JL \cdot KL(3(IL \cdot JL + 1) + 1)$$
$$\simeq 6IL^2 JL^2 K \tag{C-21}$$

or for $IL = JL = KL = N$,

$$\theta_{rep,\, B,\, 3D} \simeq 6N^5 \tag{C-22}$$

It is an advantage of any point-iterative method such as SOR that the number of iterations required, though strongly dependent on the number of nodes in any direction, is not strongly dependent on dimensionality *per se*. Thus, the 3D SOR method has an operation count of just N times the order of the 2D method, or $0(N^4 \ln N)$. Thus, Eq. (C-22) indicates that the banded GE in 3D is slower than SOR, even for repeat solutions. Eq. (C-20) indicates it is $\tfrac{4}{3} N^2$ times slower yet, for initialization. The storage required is again $(3B + 1)n$, but B is given by (C-18), giving storage = $(3(IL \cdot JL + 1) + 1)n$. For IL = JL = KL = N, this is a storage penalty of $(3N^2 + 3)$ 3D arrays, which will be insurmountable for large problems.

C.7 Conclusions

For 2D non-orthogonal but fixed grid problems with small to moderate resolution solved many times on large memory computers, banded Gaussian elimination has much to recommend it. For virtually any other class of problems, and certainly in 3-D, it is not a viable choice.

References for Appendix C

1. H. Dwyer, K. Matsuno, S. Ibrani, and M. Hafez, Some Uses of Direct Solvers in Computational Fluid Mechanics, AIAA Paper 87-0594, 25th AIAA Aerospace Sciences Meeting, Reno, NV, Jan. 1987.
2. C. P. Van Dam and M. Hafez, Comparison of Iterative and Direct Solution Methods for Viscous Flow Problems, *AIAA Journal*, Vol. 27, No. 10, Oct. 1988, pp. 1459-1461.
3. M. Hafez. S. Palaniswamy, and P. Mariani, Calculations of Transonic Flows with Shocks using Newton's Method and Direct Solver: Part II. Solution of Euler Equations, AIAA Paper 88-0226, 26th AIAA Aerospace Sciences Meeting, Reno, NV, Jan. 1988.
4. M. Hafez and D. Brucker, Unsteady Navier-Stokes Calculations Using Biharmonic Formulation and Direct Solver, AIAA Paper 89-0465, 27th AIAA Aerospace Sciences Meeting, Reno, NV, Jan. 1989.
5. R. E. Ewing, Guest Editor, *Computer Methods in Applied Mechanics and Engineering*, Vol. 47, 1984.
6. P. J. Roache, *Computational Fluid Dynamics*, rev. printing, Hermosa Publishers, Albuquerque, NM, 1976.
7. D. A. Knoll and P. R. McHugh, A Fully Implicit Direct Newton's Method for the Steady-State Navier-Stokes Equations, *International Journal for Numerical Methods in Fluids*, Vol. 17, 1993, pp. 449-461.
8. D. W. Riggins, R. W. Walters, and D. Pelletier, Effiecient Use of Direct Solvers for the Calculation of Compressible Flows, *AIAA Journal*, Vol. 29., No. 2, February 1991, pp. 311-312.
9. J. Dongarra, C. B. Moler, J. R. Bunch, and G. W. Stewart, *LINPACK Users' Guide*, Society for Industrial and Applied Mathematics, Philadelphia, 1979.
10. E. Anderson et al., *LAPACK Users' Guide*, Society for Industrial and Applied Mathematics, Philadelphia, 1992.

Index